RARE GENETIC DISORDERS
THAT AFFECT THE SKELETON

RARE GENETIC DISORDERS THAT AFFECT THE SKELETON

H.J. Mankin, MD & K.P. Mankin, MD

authorHOUSE®

AuthorHouse™ LLC
1663 Liberty Drive
Bloomington, IN 47403
www.authorhouse.com
Phone: 1-800-839-8640

Published by AuthorHouse   09/18/2013

ISBN: 978-1-4918-1501-4 (sc)
ISBN: 978-1-4918-1504-5 (e)

Library of Congress Control Number: 2013916579

This book is dedicated to Carole J. Mankin (1932-2012) – loving wife, mother and helper.

CONTENTS

CHAPTER 1

Introduction

Genetic disorders that have an effect on connective tissue and especially skeletal parts are rarely encountered by orthopaedic surgeons. The patients are sometimes mildly altered in function and structure but may also present with some severe alterations which not only cause severe disability and deformity, but also mental retardation and early death. Up to this date, there is limited knowledge as to the genetic errors, the causation of localized soft tissue and skeletal alterations and only very limited information as to possible treatment.

The purpose of introducing the data for these 14 diseases is to help caretakers and especially orthopaedic surgeons identify the patients based on their clinical presentations and also perform research to assess the causes of the diseases and possibly develop treatment protocols which may help the patients survive and improve their function.

In addition to the obvious benefits of providing a compendium of knowledge about these disorders, the book as a whole is an important examination of the benefits of careful observation and collaboration. Each of these entities was meticulously and fully described in a time before the advent of instant communication, yet to collect the cases necessary to validate each was a monumental effort. The recognition of repetitive patterns allowed the entities to be grouped together as specific syndromes. And despite the fact that the concept of genes was only barely understood in their times, the scientists who observed these diseases recognized their heritable nature. It is to their credit that each one has proven have a definite genetic basis.

As medicine enters the age of the genome, diseases like the ones described here—progressive and debilitating, but potentially reversible—would seem to be a frontier for gene therapy. So despite being a catalog of unusual syndromes, this book may actually contain the seeds of a medical revolution.

CHAPTER 2

Alkaptonuric Ochronosis

One of the rarest and most unusual genetic disorders is alkaptonuric ochronosis; a disease that sometimes spectacularly affects bones, joints, kidneys, eyes, skin and heart. The disease was described centuries ago and has been found in most countries of the world, some with greater frequency. It is a believed to be caused by a genetic error which becomes evident in patients in their mid years. The urine of the affected patients turns black on exposure to oxygen, and the patients develop black skin and eye discolorations, sometimes life threatening aortic and cardiac valve disease, and often very severe spinal and joint disorders. The biochemical cause of the disorder was discovered to be the defective activity of the enzyme homogentisic acid oxidase; and it was the first entity in history to be considered to be a genetic error. In that respect it became a model for other disorders. The problems are that the gene locus has never been completely identified; and at least until recently no reasonable approach has been established to treat the patients and prevent these often devastating changes.

Nomenclature and history:

Alkaptonuric ochronosis is also know as Garrod's disease, ochronosis, alcaptonuria, homogentisuria, and "dark urine disease" (6). There is some modest confusion since both the words "alkaptonuria" (German) and "alcaptonuria" (French) appear in the literature. The term describes the dark coloration of body fluids. The term "ochronosis" relates to the "ochre" coloration of connective tissues, especially collagen. Most authors and clinicians use the term "alkapatonuric ochronosis" to describe the disease.

In a remarkably comprehensive review, Garrod (31) stated that this disease was first noted by G. A. Scribonius in 1584, then by Schenck in 1609 and then by Zacutus Lusitanus in 1649. It was not until 1859 however that Boedeker first recognized the peculiar color of the urine and named it "alkaptonurie"(8). The morbid anatomy was first described by Virchow in 1866 (74), who reported an ochre coloration in the resected tissues and introduced the term "ochronosis". It was not until Albrecht's studies in 1902 (2) and Osler's description in 1904 (60), that the two words involved in the disease were put together to name it "alkaptonuric ochronosis". The abnormal material was identified in 1891 by Wolkow and Baumann (76) as 2,5-dihydroxyphenylacetic acid and they renamed it as homogentisic acid. In 1909, Neubauer mapped the tyrosine degradation process (57), which was subsequently established by La Du and coworkers as the major factor in the development of alkaptonuria (43). Severe arthritis was first reported by Gross and Allard in 1907 (33) and the arthritic changes defined by Söderbergh in 1913 (68). In 1932, Hogpen and associates (34) reviewed large numbers of cases and concluded that the genetic error was a recessive one, a factor which was subsequently at least partially challenged by Milch in 1955 (54). Of interest is the finding of alkaptonuric ochronosis in an Egyptian mummy described by Stenn in 1977 (71) and the presence of the disorder in some animals (6).

Sir Archibald Garrod, in a series of signal contributions (29-31) extended the knowledge of the disorder and on the basis of his studies of alkaptonuric ochronosis, he defined the nature of all genetic disorders . . . using the phrase "one gene-one enzyme". He wrote a famous book on genetic disorders in 1909 (31) in which he stated "Of inborn errors of metabolism, alkaptonuria is that of which we know the most and from the study of which the most has been learned".

Pathogenesis of alkaptonuric ochronosis:

Alkaptonuria is a rare autosomal recessive disorder caused by a deficiency in homogentisate-1,2-dioxygenase activity, which leads to accumulation of large amounts of homogentisic acid in the body (6,12,20,32,39,43,52,53,62). The material produced is bound to collagen and other materials including osseous, cartilaginous, cardiac, renal and ocular structures as an oxidized and polymerized dark pigment (called "ochronosis") (6,9,12,16,39,47,53,62,79). Homogentisic acid itself is excreted in the urine and turns dark brown or black on oxygenation and alkalinization. It should be noted that ochronosis is probably not due to the presence of homogentisic acid alone but rather its oxidation product, benzoquinone acetic acid, which binds either directly or after polymerization to biologic materials and especially collagen (3,6,12,21,24,28,32,39,43,62,79).

There is very limited information regarding the genetic error in patients with alkaptonuric ochronosis. Janocha et al in 1994 (37) reported that the human gene for alkaptonuria mapped to chromosome 3q13 and since then, a large number of mutations have been described which may account for the variability of the presentations. The mutations are mostly missense and are distributed throughout the HGD sequence (homogentisate1,2 dioxygenase, which is 54.3 kb in length and has 14 exons). (24,32,39,62,63, 65,66,80). The disorder is thought to be an autosomal recessive error but a study by Milch in 1957 (54) seems to support the possibility that the disorder is sometimes dominant. It is evident that there are some populations, who seem to have a much higher frequency. These include persons from Slovania, Czechoslovakia and the Dominican Republic but some of that may be related to the frequency of close relative marriages in those communities (1,6,24,58). The disease has been described as occurring in Japanese, Indians, Italians, African-Americans, Turks and many other populations (6). It is slightly more common in males than females. The disorder is often diagnostic at birth partly because of the finding of "black diapers", but skin changes, ear problems, renal disorders, vascular disturbances and bone and joint disease usually do not present until after the age of 30 (6,10,13,39, 47,59,62).

Histologic and imaging abnormalities:

Elevated levels of homogentisic acid enter all tissues and with oxidation, display black pigmentation (6,9,12,21,28,49,59,62). A polymerized oxidized form of the material is deposited in various types of tissues binding irreversibly to collagen and causing bluish-black pigmentation (ochronosis) (6,9,12,21,28,39,59,62). The material and resultant color change can be seen histologically in articular and meniscal cartilage, tendons, ligaments, joint capsules, sclerae, larynx, bronchial tissue and cardiac valves. Cartilage is the most darkly pigmented tissue in joints while capsular and tendinous tissues are much less affected (5,9,44,49,62). The discal tissues in the spine are severely affected and have extensive fibrotic and ultimately ossific changes which result in a fused and markedly deformed spine. Osteophytes are almost always present (6,28,45,51,79). In the kidney and blood vessels, granules of pigment may be seen in tubular and vascular channels and valves may be green in color. In the prostate, the entire structure shows ochronosis with pigmented calculi (6,10,23,28,35,78,81).

Imaging studies are often very striking. Calcification of soft tissues is sometimes subtle with the exception of the prostate, which may be almost entirely calcified (22,35,38, 41,78). Calcification of renal tissue is also common. The spine has the most striking changes with narrowing of the intervertebral spaces and calcification of the disks, which present as a spindle shape on lateral projection (6,38,64). Sometimes the disk space has a vacuum appearance and the spine may be markedly shortened. In severely involved cases, osteophytes lead to fusion of the lumbar, thoracic and even cervical spine and resultant deformity (17,55). Major joints are often markedly altered. The changes are those of severe osteoarthritis with narrowing of the joint space, osteophytes and irregularity of bony contour (6,48). A remarkable degree of calcification may be present particularly in the cartilages of the knee and shoulder joint and as loose bodies within the joint. Pubic synchondrosis occurs in almost all patients (6). Special imaging studies of the heart and aorta will sometimes show severe changes associated with aortic stenosis and valvular disease (41,42).

Clinical Presentation:

Children with alkaptonuric ochronosis usually have normal mentation and show no significant physical abnormalities, with one major exception . . . the syndrome of "black urine". The children grow normally and are functionally able but their urine if it becomes exposed to air or become alkalinized turns black and the syndrome of "black diapers" causes considerable concern to parents and caring physicians (6,62). Once adulthood is reached however the syndrome has multiple presentations (6,39,41,62).

1. *Deposits in eyes, ears and skin*: Scleral darkening and sometimes markedly black conjunctival nodules can be noted at the age of 20 or less in affected patients (4,6,13,14,47,61,69,73) (Figure 2). Visual disturbances may occur as a result of the conjunctival deposits. Ear lesions are located in the soft tissues and skin of the ears, which may become slate gray or blue in color and sometimes show black nodules (6,59,61,62). Usually there is no hearing loss in association with these changes. Changes in skin color may be present in the digits of the hands and feet or even in the forearm and legs (6,47,62,73). They are

usually asymptomatic. Skin sweat may turn black as it is exposed to air and although not symptomatic, causes problems with clothing (62).

2. *Prostatic and renal calculi*: Prostatic and renal calculi usually start to appear in middle years and can be devastating in terms of development of symptoms, especially related to passage of urine. Although renal obstructive problems may occur, the patients only rarely develop renal failure and most often can be treated effectively by increasing the fluid level (6,36,62,78). The prostate gland is often completely involved and prostatectomy may be necessary to prevent urinary obstruction (78).

3. *Bone and joint disorders*: Most of these begin in the fourth decade and principally affect the spine and major joints (1,3,9,12,17,22,44,45,49,51,59). The spinal changes are quite striking with marked narrowing of the intervertebral disk spaces and ultimately complete collapse and calcification of the spaces so as to fuse the spine, often in marked scoliosis and kyphosis (17,45,51,55,64). Joint spaces are markedly diminished in the knees, hips and shoulders and calcification of cartilage, menisci and surrounding soft tissues may occur (1,6,9,18,25,38,44). The joints often become severely arthritic and the patients become disabled by the time they reach the age of 50. The pubic symphysis is frequently affected. The tendons in the lower extremity are often severely involved especially the Achilles tendon, which may become markedly thickened, calcified and occasionally ruptures (46,50).

4. *Cardiovascular abnormalities:* These are clearly the most serious problems and in some cases life-threatening. Vascular abnormalities including arteriosclerosis and aortic stenosis often appear in older patients (10,11,19,23,25,62,81). The abnormalities principally affect the aortic and mitral valves and may cause severe vascular compromise (10,11,23,25,62,81). Myocardial infarction may occur and is the principal cause of death for these patients in their 60s and beyond.

Treatment of alkaptonuric ochronosis:

Over many years, an array of agents has been introduced to treat alkaptonuric ochronosis but none have proved to be very effective. They include vitamins, tyrosinase, insulin, adrenal cortical extract, corticosteroids and thiouracil (6,7,12,26,62,75,77). Early in the course it was proposed that ascorbic acid may have an effect on alkaptonuria and studies performed by a number of physicians seemed to suggest that high doses decreased the binding of homogentisic acid into collagenous structures (27,56,67,77). Although the patients did not seem to improve, high doses of Vitamin C caused a decrease in benzoquinone acetic acid, which is the binding agent involved with collagen (67,77).

Another approach has been dietary manipulation. A diet low in phenylalanine and tyrosine seems to reduce the urinary homogentisic acid (6,56). The effects were minimal and most people found it difficult to tolerate the diet. One case of "cure" occurred in a patient who received a liver transplant for hepatitis B cirrhosis (40) and another showed vast improvement after renal transplantation (36). Neither of these methods seem reasonable. Finally, in recent years a new agent, nitisisone, a potent inhibitor of 4-hydroxyphenolpyruvate dioxygenase has been introduced and at least in very short exposure seems to reduce production and urinary excretion of homogentisic acid. The treated patients seem to do well but the complication rate has not as yet been fully established (62,72).

Surgeries for prostatic and renal cell problems are often helpful and aortic and valvular surgery is sometimes required to save the patient's life (10,11,19,23,25,42). Finally it should be evident that orthopaedic problems often require major treatment. Arthroscopic "washout" of affected joints has not been other than transiently successful (48). Total joint replacements are quite frequently required in these patients for severe disorders in their hips, knees and shoulders and are generally successful (6,9,15,18,25,44,70). Dealing with the spine is a major issue and unless the spinal cord is damaged, the surgery is not only difficult but has not resulted in great improvement (45,51,55).

Conclusion:

There is little doubt that alkaptonuric ochronosis is one of the most unusual genetic disorders of humans and certainly it was one of the earliest to be described and biochemically defined. The presence of black urine early in life causes great consternation and the development of orthopaedic, cardiac, ocular and renal afflictions with advancing age make it dramatic and at times, debilitating. Scientists fully understand the cause of the symptoms and findings for the afflicted patients and can follow them with serial studies and at times perform surgery to make life more acceptable. The problems with the disorder are two fold, however. The first is that despite very close to a century of identification and understanding of the mechanism of production of homogentisic acid, we have not really clearly identified the genetic error that causes it. The second is that a large number of treatments have been tried and we still have not solved a way to diminish the effect of the biochemical abnormalities on the patient. Surgical approaches help the patients with their heart, kidney and especially their joint problems, but don't prevent or alleviate some of their disability. We need to know more about this disease and try to figure out how some biologic solutions can restore the patients to good health.

References:

1. Abe YH, Oshima N, Hatanaka K et al: Thirteen cases of alkaptonuria from one family tree with special reference to osteo-arthrosis alkaptonurica. J Bone Joint Surg 40A; 817-831, 1960.
2. Albrecht H: Ueber ochronose. Z Heil Path Anat 23: 366-378, 1902.
3. Aliberti G, Pulignano I, Schiappoli A et al: Bone metabolism in ochronotic patients. J Intern Med 254-296-300, 2003.
4. Allen RA, O'Malley C, Straatsma BR: Ocular findings in hereditary ochronosis. Arch Opthalmol 65: 657-668, 1961.
5. Bauer J, Kienbock R: Zur kentniss der knochen-und gelenksveranderungen bei alkaptonurie. Osteoarthrosis alcaptonurica (ochronotica) Fortschr Geb Roentgenstrahlen 40: 32-42, 1929.
6. Beighton P, Berman P, Srsen S: Alkaptonuria. In McKusick's Heritable Disorders of Connective Tissue. 5th Edition. Beighton P, Ed. St. Louis, Mosby 315-334. 1993.
7. Black RL: Use of cortisone in alkaptonuria. J Am Med Assoc 155:968-970, 1954.
8. Boedeker C: Ueber das alcapton ein neuer beitrag zur frage: welche stoffe des harns können kupferreduction bewirken? Z Ration Med 7: 130-145, 1859.
9. Borman P, Bodur H, Ciliz D: Ochronotic arthropathy. Rheumatol Int 21: 205-209, 2002.
10. Butany JW, Naseemuddin A, Moshkowitz Y, Nair V: Ochronosis and aortic valve stenosis. J Card Surg. 21: 182-184, 2006.

11. Cercek M, Prokselj K, Kozelj M: Aortic valve stenosis in alkaptonuric ochronosis. J Heart Valve Dis. 11: 386-388, 2002.

12. Cervenansky J, Sitaj S, Urbanek T: Alkaptonuria and ochronosis. J Bone Joint Surg 41A: 1169-1182, 1959.

13. Cheskes J, Buettner H: Ocular manifestations of alkaptonuric ochronosis. Arch Opthalmol. 118: 724-725, 2000.

14. Cheves Barrios P, Font RL: Pigmented conjunctival lesions as initial manifestation of ochronosis. Arch Opthalmol 122: 1060-1063, 2004.

15. Cottini E: Hip prosthesis in alkaptonuric arthropathy of familial type. Clin Orthop 25: 44-59, 1974.

16. Crissey RE, Day AJ: Ochronosis: a case report. J Bone Joint Surg 32A: 688-690, 1950.

17. Deeb Z, Frayha RA: Multiple vacuum discs: an early sign of ochronosis: radiologic findings in two brothers. J Rheumatol 3: 82-87, 1976.

18. Demir S: Alkaptonuric ochronosis : a case with multiple joint replacement arthroplasties. Clin Rheumatoil 22: 437-439, 2003.

19. Dereymaeker L, Van Parijs G. Bayart M et al: Ochronosis and alkaptonuria: report of a new case with calcified aortic valve stenosis. Acta Cardiol 45: 87-92, 1990.

20. Deutsch JC, Santhosh-Kumar CR: Quantitation of homogentisic acid in normal human plasma. J Chromatogr B Biomed Appl 677: 147-151, 1996.

21. Di Franco M, Coari G, Bolnucci E: A morphological study of bone and cartilage in ochronosis. Virchow's Arch 436: 74-81, 2000

22. Ehrlich L: Ein Fall von ochronotischer alkaptonuric mit erheblichen skelettveränderungen. Röntgenpraxis 4: 865-870, 1932

23. Erek E, Casselman FPA, Vanermen H: Cardiac ochronosis: valvular heart disease with dark green discoloration of the leaflets. Tex Heart Inst J. 31: 445-447, 2004.

24. Fernandez-Canon JM, Granadino B, Beltran-Valero de Parnabe D et al: The molecular basis for alkaptonuria. Nat Gent 14: 19-24, 1996.

25. Fisher AA, Davis MW: Alkaptonuric ochronosis with aortic valve and joint replacements and femoral fracture: a case report and literature review. Clin Med Res 2: 209-215,2004.

26. Flaschenträger B, Halawani A, Nabeh I: Alkaptonurie and vitamin B12. Klin Wschr 32:131-133, 1954.

27. Forslind K. Wollheim FA. Akesson B. Rydholm U: Alkaptonuria and ochronosis in three siblings: ascorbic acid treatment monitored by urinary HGA excretion. *Clin Exp Rheumatol. 6(3):289-92.*

28. Gaines JJ Jr. the pathology of alkaptonuric ochronosis. Hum Pathol. 20: 40-46, 1989.

29. Garrod AE: About alkaptonuria. Lancet 2: 1484-1486, 1901

30. Garrod AE: The incidence of alkaptonuria: a study in chemical individuality. Lancet 2: 1616-1620, 1902

31. Garrod AE: Inborn Errors of Metabolism. London; Frowde, Hodden and Stoughton; 1909.

32. Granadino B, Beltran-Valero de Bernabe D, Fernandez-Canon JM et al: The human homogentisate 1,2,-dioxygenase (HGO) gene. Genomics. 43: 115-122, 1997.

33. Gross O, Allard E: Unterzuchungen über alkaptonurie. Z Klin Med. 64: 359-369, 1907.

34. Hogben L, Worrall RL, Zieve L: The genetic basis of alkaptonuria. Proc Roy Soc Edinb 52: 264-295, 1932

35. Hollingsworth RP: Homogentisic acid stone: a rare form of prostatic obstruction. Br J Urol. 40: 546-547, 1968.

36. Introne WJ, Phornphutkul C, Bernardini I et al: Exacerbation of the ochronosis of alkaptonuria due to renal insufficiency and improvement after renal transplantation, Mol Genet Metab 77: 136-142, 2002.

37. Janocha S, Wolz W, Srsen Set al: The human gene for alkaptonuria (AKU) maps to chromosome 3q. Genomics 19: 5-8, 1994.

38. Justesen P, Andersen P: Radiological manifestations in alkaptonuria. Skeletal Radiol 11: 204-208, 1984.

39. Keller JM, Macaulay W, Nercessian OA, Jaffe IA: New developments in ochronosis: review of the literature. Rheumatol Int 25: 81-85, 2005.

40. Kobak AC, Oder G, Kobak S et al: Ochronotic arthropathy: disappearance of alkaptonuria after liver transplantation for hepatitis B-related cirrhosis. J Clin Rhematol 11: 323-325, 2005.

41. Kocyigit H, Gurgan A, Terzioglu R, Gurgan U: Clinical, radiographic and echocardiographic findings in a patient with ochronosis. Clin Rheumatol 17: 403-406, 1998.

42. Kragel AH, Lapa JA, Roberts WC: Cardiovascular findings in alkaptonuric ochronosis. Am Heart J 120: 1460-1463, 1990.

43. LaDu BN, Zannoni VG, Laster L, Seegmiller JE: The nature of the defect in tyrosine metabolism in alcaptonuria. J Biol Chem. 230: 251-260, 1951.

44. Laskar FH, Sargison D: Ochronotic arthropathy: a review with four case reports. J Bone Joint Surg 52B: 653-666, 1970.

45. Liu W, Prayson RA: Dura mater involvement in ochronosis. Arch Pathol Lab Med 125: 961-963, 2001.

46. Loogroscino CA, Sacchettoni Logroscino G, Capoale M: Subcutaneous Achilles tendon rupture in alkaptonuria: clinical contribution: microscopic and ultrastructural features. Arch Putti Chir Organi Mov 32: 307-311, 1982.

47. Lubics A, Schneider I, Sebok B, Havass Z: Extensive bluish gray skin pigmentation and severe arthropathy. Endogenous ochronosis (alkaptonuria). Arch Dermatol 136: 548-549, 2000.

48. Lurie DP, Musil D: Knee arthropathy in ochronosis. Diagnosis by arthroscopy with ultra-structural features. J Rheumatol 11: 101-103, 1984.

49. Mannoni A, Selvi E, Lorenzini S et al: Alkaptonuria, ochronosis and ochronotic arthropathy. Semin Arthritis Rheum. 33: 239-248, 2004.

50. Manoj Kumar RV, Rajasekaran S: Spontaneous tendon ruptures in alkaptonuria. J Bone Joint Surg 85B: 883-886, 2003.

51. McCollum DE, Odom GL: Alkaptonuria, ochronosis and low back pain. J Bone Joint Surg. 47A: 1389-1392, 1965.

52. Milch H, Milch RA: Alcaptonuria. J Int Coll Surg 15: 669-685, 1951.

53. Milch RA: Biochemical studies on the pathogenesis of collagen tissue changes in alcaptonuria. Clin Orthop 24: 213-229, 1962.

54. Milch RA, Milch H: Dominant inheritance of alcaptonuria. Acta Gent 7:178-184, 1957.

55. Millea TP, Segal S, Liss RG, Stauffer ES: Spine fracture in ochronosis. Report of a case. Clin Orthop 208-211, 1992.

56. Morava E, Kosztolanyi G, Engelke UF, Wevers RA: Reversal of clinical symptoms and radiographic abnormalities with protein restrictions and ascorbic acid in alkaptonuria. Ann Clin Biochem 40: 108-11, 2003.

57. Neubauer O: Uber den abbau der aminsauren im gesunden und kranken organismus. Dtsch Arch Klin Med 95: 211-256, 1909.

58. Neuwirth A, Srsen S. Petricek J et al: Analysis of alkaptonuria incidence in one region of northwest Slovakia: a preliminary report. Birth Defects. 10: 244-249, 1974.

59. O'Brien WM, La Du DN, Bunim JJ: Biochemistry, pathologic and clinical aspects of alcaptonuria, ochronosis and ochronotic arthropathy. Am J Med 34: 813-838, 1963.

60. Osler W: Ochronosis: the pigmentation of cartilages, sclerotics and skin in alkaptonuria. Lancet 1:10-11, 1904.

61. Pau HW: Involvement of the tympanic membrane and ear ossicle system in ochronotic alkaptonuria. Laryngol Rhinol Otolaryngol 63: 541-544, 1984.

62. Phornphutkul C, Introne WJ, Perry MB et al: Natural history of alkaptonuria. N Engl J Med 347: 2111-2121, 2002.

63. Pollok MR, Chou YH, Cerda JJ et al: Homozygosity mapping of the gene for alkaptonuria to chromosome 3q2. Nat Genet. 5: 201-204, 1993.

64. Pomeranz MM, Friedman LJ, Tunick IS: Roentgen findings in alkaptonuric ochronosis Radiology 37: 295-303, 1941.

65. Porfirio B, Chiarelli I, Graziano C et al: Alkaptonuria in Italy: polymorphic haplotype background mutational profile and description of four novel mutations in the homgentisate 1-2-dioxygenase gene. J Med Genet 37:309-312, 2000.

66. Rodriguez JM, Timm DE, Titus GP et al: Structural and functional analysis of mutations in alkaptonuria. Hum Mol Genet 9: 2341-2350, 2000.

67. Sealock RR, Galdston M, Steele JM: Administration of ascorbic acid to an alkapatonuric patient. Proc Soc Exp Biol Med 44: 580-583, 1940

68. Söderbergh G:Ueber ostitis deformans ochronotica. Neur Zbl 32: 1362-1363, 1913.

69. Smith JW: Ochronosis of the sclera and cornea complicating alkaptonuria: review of the literature and report of four cases. J Am Med Assn. 120: 1282-1288, 1942.

70. Spencer JMF, Gibbons CLMH, Sharp RJ et al: Arthroplasty of ochronotic arthritis. No failure of 11 replacements in 3 patients followed 6-12 years. Acta Orthop Scand 75: 355-358, 2004.

71. Stenn FF, Milgram JW, Lee SL et al: Biochemical identification of homogentisic acid pigment in an ochronotic Egyptian mummy. Science 197: 566-568, 1977.

72. Suwannarat P, O'Brien K, Perry MB et al: Use of nitisinone in patients with alkaptonuria. Metabolism 54: 719-728, 2005.

73. Vijaikumar M, Thappa DM, Srikanth S et al: Alkaptonuric ochronosis presenting as palmoplantar pigmentation. Clin Exp Dermatol 25: 305-307, 2000.

74. Virchow R: Ein fall von allgemeiner ochronose der knorpel und knorpelahnlichen theile. Virchow's Arch Pathol Anat. 37: 212-219, 1866.

75. White AG, Parker J, Block F: Studies in human alkaptonuria: effect of thiouracil, para-aminobenzoic acid and di-iodotyrosine on excretion of homogentisic acid. J Clin Invest 38: 140143, 1949.

76. Wolkow M, Baumann E: Ueber das wesen der alkapatonurie. Z Physiol Chem 15: 228-285, 1890-1891.

77. Wolff JA, Barshop B, Nyhan WL et al: Effects of ascorbic acid in alkaptonuria: alterations in benzoquinone acetic acid and an ontogenic effect in infancy. Pediatr Res 26: 140144, 1989.

78. Young HH: Calculi of the prostate associated with ochronosis and alkaptonuria. J Urol 51:48-58, 1944.

79. Zannoni VG, Lomtevas N, Goldfinger S: Oxidation of homogentisic acid to ochronotic pigment in connective tissue. Biochem Biophys Acta 177: 94-105, 1969.

80. Zatkova A, Chmelikova A, Polakavoa H et al: Rapid detection methods for HGO gene mutations causing alkaptonuria. Clin Genet 63: 145-149. 2003.
81. Zund G, Schmid AC, Vogt PR et al: Green aortic valve: alcaptonuria (ochronosis) with severe aortic stenosis Ann Thorac Surg. 67: 1805, 1999.

CHAPTER 3

Homocystinuria: A Complex Genetic Disorder

Introduction:

Homocystinuria is a rare genetic disorder first described in the Irish population in the early 1960's. The disease appears to be caused by a deficiency of cystathionine beta synthase (CBS) and results in increased amounts of homocysteine and methionine in the blood and a marked increase in homocystine in the urine of affected patients. The clinical presentation includes mental retardation, seizures, psychiatric problems, increased body height, premature arteriosclerosis and thrombosis, ocular disturbances and sometimes, skeletal abnormalities with severe osteoporosis. In some patients, the disease can be treated effectively by administration of pyridoxine (Vitamin B6).

Nomenclature and history:

Alternative titles for homocystinuria include cystathionine-beta deficiency, CBS deficiency, homcystinemia, methionine metabolic disorder.

The original description of the disorder now known as homocystinuria was by Carson and Neill in 1962 (9) who screened some mentally retarded patients in Ireland and discovered that their urine contains large amounts of homocystine. Some of these patients were "Marfanoid", in terms of being very tall, but were also mentally retarded and lacked the loose structure to their joints and especially their hands. Furthermore some of the patients had ectopia lentis (dislocating lens). Almost simultaneously Gerritsen and coworkers in Madison, Wisconsin (20,21) described the same metabolic defect in a child with cerebral palsy and also identified the presence of increased concentrations of homocystine in the patient's urine. Shortly thereafter in 1963, Komrower and Wilson (31) presented a signal paper at the Royal Society of Medicine which introduced the name "homocystinuria" for the disease. In 1964, Gibson, Carson and Niell (22) described thromboembolic phenomena in some affected patients. In 1965, Schimke, Pollack and Victor McKusick at John's Hopkins Hospital (60) described 38 cases in 20 families and in the following year, McKusick, in his outstanding book on heritable connective tissue disorders

11

(43) further defined the sometimes subtle differences from Marfan's syndrome. Chou and Waisman in 1965 (11) performed an autopsy on a patient and described "spongy degeneration" of the nervous system. In 1966, Dunn and coworkers (15) further evaluated the nature of the neural problems and in the same year, a study by Bett (7) pointed out that some of the patients with homocystinuria had epilepsy. Mudd and his coworkers in the early 1960s (46,47) became interested in the disease and not only defined the nature of the disorder but that it appeared to be caused by genetic error in the production of cystathionine-beta-synthase. In 1967, Presley (52) described the ocular defects, Brenton the problems related to diet (5) and Smith in an article in American Journal of Roentgenology in the same year (65) showed illustrations that defined the sometimes unique changes in the skeleton. In 1969, Beals in the Journal of Bone and Joint Surgery defined aspects of the skeletal problems (3).

Biologic and genetic characteristics of homocystinuria:

The cardinal biologic characteristics of homocystinuria are markedly increased concentrations of plasma homocystine, homocysteine and methionine; increased urine concentrations of homocystine; and reduced cystathionine beta-synthase enzyme activity (18,28,33,36,44,48,54,66,67,74). These changes appear to affect the collagen, fibrillin-1 and other support structural elements of the bones, blood vessels, heart and eyes (30,41,42,54,59,69). The neural elements are also affected and this is most obvious in terms of mental retardation and neurologic symptomatology, some of which are seriously disabling (4,7,11,16,40,58,63,68). These latter events appear to result from increased serum concentrations of homocysteine, which also seem closely related to the development of osteoporosis (27).

The disorder is inherited as an autosomal recessive trait and is somewhat more frequent in males than in females (43,54). As indicated in the early studies the frequency in the population of northern Ireland (1 in 65,000) exceeds that of other world sites (approximately 1 in 350,000) (9,18,38,50,54,74).

Homocystinuria has multiple forms and expression, which may be in part based on the variation in genetic origin (18,32,38,54,67). Some patients are only mildly affected and have a relatively normal life expectancy and others present with extensive neurologic, skeletal and cardiovascular abnormalities associated with a much shorter life span. The gene map locus is 21q22.3 (54). The error appears to be a deficiency of cystathionine beta synthase (CBS), which leads to a progressive increase in the synthesis of methionine and homocystine, the latter of which is especially noted in the urine (18,28,33,36,44, 48,54,66,67,74). There are at least 90 mutations which have been identified in many laboratories and most are missense (18,28,32,36,44,54). The two most frequently encountered mutations are I278T, which is sensitive to administration of pyridoxine and hence is not only easier to treat but milder in presentation (32,38,44,54). The G307S mutation is most common particularly in Celtic individuals. This system is not pyridoxine (Vitamin B6) sensitive and hence can be more severe and potentially damaging to the patient (18,44,54,74).

Another confusing issue regarding this disorder is that there are a number of ways to cause hypocystinuria, some of which are less closely related to cystathionine beta-synthase production (26,32,36,37,44,47,54,66). An additional seven rarer causes for the homocystinuric state include:

1. Defect in Vitamin B12 metabolism.
2. Deficiency of N(5,10)-methlyenetetrahydrofolate reductase.
3. Selective intestinal malabsorption of Vitamin B12.
4. Vitamin B12-responsive homocystinuria cbl E type.
5. Methylcobalamin deficiency cbl G type.
6. Vitamin B12 metabolic defect, type 2.
7. Transcobalamin II deficiency.

Clinical presentation of patients with homocystinuria:

The clinical findings for patients with homocystinuria include central nervous problems, skeletal features, ocular disturbances, cardiovascular abnormalities and changes in the skin (28,43,54). It is important to note that patients with homocystinuria resemble those with Marfan's syndrome (6,23,54). Both syndromes show increased height, dolichostenomelia, arachnodactyly osteoporosis, ectopia lentis, cardiac disease and spinal deformities. There are however some major differences that distinguish the two diseases. First and foremost, most of the homocystinuric patients are not only mentally retarded but have an array of central nervous system abnormalities, which are rarely encountered in the Marfan's group (1,6). In the the ectopia lentis for homocystinuria, the lens is dislocated downward, toward the cheek while for Marfans the lens goes upward toward the eye-brow (13). Foot deformities are far less common or disabling for homocystinura and the joints tend to be stiffer in both the hand and foot. Hypermobility such as is seen with the Steinberg thumb sign and the Walker-Murdoch wrist sign, both of which are virtually diagnostic for patients with Marfan's syndrome (70) are rarely seen in homocystinuria. The critical tests which distinguish the two maladies are for homocystinuria, the markedly increased concentrations of plasma homocystine, homocysteine and methionine and the especially the increased urine concentrations of homocystine, none of which occur in Marfan's syndrome (2,10,17,20,51,52,54).

Central nervous system problems for patients with homocystinuria include first and most devastating, the frequent occurrence of mental retardation, which can be noted quite early in the child's life (1,43,54,75). It is more severe in the pyridoxine insensitive group even when treated, and progresses with advancing years (75). Well over 50% of the entire series of patients have IQs of less than 60 (43,54). Seizures of the grand mal type occur in over 20% of the untreated patients and even without the actual events, EEGs are often positive (7,54). Dystonia is common and personality disturbances including depression and schizophrenia are quite common especially in the severe form of the disease (1,16,40,58,63). In the pyridoxine insensitive form of the disease, intracranial venous and arterial thromboses may cause headache, neck stiffness and bloody cerebrospinal fluid (54,68). The calvarium is usually small in size on X-ray examination (54).

Ocular problems are commonly seen. Myopia is the most common problem, which in a high percentage of the patients is followed by ectopia lentis, with the lens displaced downward toward the cheek (8,13,19,52). Glaucoma is common, as is buphthalmos, retinal detachment and optic atrophy, particularly in patients who require surgery for dislocated lenses (25,29,49,54,57).

Skeletal disorders are among the most striking and disabling aspects of the disease. Virtually all of the patients are excessively tall in height even when treated and have very thin extremities and reduced body weight (3,43,48,54,65,67). Curvature of the spine and platyspondyly reduce the overall height somewhat, particularly as the patients grow older (43,54). Almost all of the patients have an early onset osteoporosis, which frequently results in fractures occurring with minor injury. This finding seems to be related to the increased level of homocysteine in the serum (54,69). Spondylolisthesis can occur and chronic back pain is a common complaint (54). Varus deformity of the proximal humerus is frequent and often results in limitation of shoulder movement (54). Flaring of the distal femur and proximal tibia are common and result in large knobby knees, sometimes limited in movement (43). In children, distal radial and ulnar physes contain tiny calcific spicules on radiographs (3,54). One of the key distinguishing features from Marfan's disorder is the absence of relaxed musculature and excessive movement and displacement of joints (6,23,54). The joints are often tight and the foot and hand are sometimes limited in movement (54).

Vascular disturbances are quite common particularly as the child ages. Thromboses can occur in many veins and arteries. Thrombophlebitis is common and starts to occur after the age of 20 (24,43,45,54,55,71). The most significant vascular occurrence is coronary occlusion, which is a common cause of death for patients with homocystinuria even when treated with pyridoxine (24,53,71,72,73). Acute gangrene particularly of the lower extremity sometimes occurs in older patients and varicose ulcers are common (53,72). Gastrointestinal bleeding is an occasional problem and can be threatening (54,73). Pulmonary embolism and pancreatitis can occur as well and both are difficult to treat (12,54).

Skin problems include malar flushing, wide pores of the facial skin, creases in the fingers and light colored hair (53,54,62). Premature graying of the hair may occur in teenagers and eczema is fairly common in patients with homocystinuria, especially those with the more severe form (62).

Management of patients with homocystinuria:

One of the critical features in dealing with patients with homocystinuria is establishing the diagnosis as quickly as possible by both spot tests for homocysteine in the blood and analysis of urine samples for homocystine (2,10, 17,50,51,54,56). A diet low in methionine has been proposed for patients with homocystinuria and in a study presented in 1985 by Mudd et al, (48) such a program appeared to reduce the frequency of ectopia lentis and seizures. For patients who have pyridoxine (vitamin B6) sensitivity, the use of this material for children at a concentration of 500 mg per day may be valuable and once symptomatology is diminished, can be reduced to 200 (54). Unfortunately, high dose pyridoxine treatment may have some life-threatening effects in infants (64). Folic acid is also valuable particularly for patients who are not pyridoxine sensitive and can reduce the homocystine concentrations in the urine (54). The combination of pyridoxine and folic acid appears to be even more successful for pyridoxine insensitive patients (54). Another agent, which appears to be successful for many patients is betaine and this may be added to pyridoxine for patients with insensitive disease (34,61). One of the problems however is the frequency of cerebral edema with betaine therapy (14). Cystine supplement may also be helpful for severe cases (35)

Surgical management for ectopia lentis is important and should be done early in the course to prevent retinal damage (8,19,25,29). Problems with the vascular system can be very serious

(45,54,55,72,73). Most patients should be treated regularly with aspirin to reduce the likelihood of clots and cardiac problems. They may also require treatment with anticoagulant elements and sometimes stent implantation (24,54,71,72,73). Vascular and especially cardiac surgery is fraught with danger not only from the operative procedure but from the hazards of anesthesia.

Pregnancy in affected females is often high risk to both the fetus and the mother (39). Patients should be carefully observed and maintained on pyridoxine and other agents to reduce the likelihood of disasters.

There are still no suggestions for approaches to the management of sometimes severe osteoporosis (54,69), Bisphosphonates could be tried but there may be too many hazards in these patients for standard doses. Calcium and Vitamin D supplements are unlikely to be of great value. Scoliosis, vertebral collapse and pathologic fractures must be treated as carefully as possible, recognizing that these patients are not good anesthetic risks (43,54).

Psychiatric care is sometimes required for these patients especially if they become depressed and potentially suicidal. Family discussions may be helpful in providing the patients with some form of support (1)

Conclusions:

With the exception of the northern Irish population, homocystinuria is a very rare disease in the world. It is transmitted as an autosomal recessive and is moderately less frequent in females. In many ways, homocystinuria resembles a more ommon disorder, Marfan's syndrome, but can be distinguished by measuring the amount of homocystine in the urine or homocysteine in the blood. There are several forms of the disease, but the two principal types are a severe form, which is insensitive to pyridoxine and a milder form, which can often be effectively treated by pyridoxine and other agents. Ocular disturbances, mental retardation, depression and vascular abnormalities make management of these patients sometimes very difficult.

The orthopaedic problems are less severe than for patients with Marfan's disease but are much more difficult to treat, chiefly because of multiple complications that can occur.

The ideal approach in the future to this problem is either to modify the gene structure to eliminate the disease before it develops; or perhaps to introduce early in the course a corrected form of cystathionine beta-synthase. That approach is being tested but as yet shows only limited success in reducing the problems for the patient, the family or the treating physicians.

References:

1. Abbott MH, Fostein SE, Abbey H, Pyeritz RE: Psychiatric manifestations of homocystinuria due to cystathionine beta synthase deficiency: prevalence, natural history, and relationship to neurologic impairment and vitamin B6 responsiveness. Am J Med Genet. 26:959-969, 1987.

2. Accinni R, Campolo J, Parolini M et al: Newborn screening of homocystinuria: quantitative analysis of total homocyst(e)ine on dried blood spot by liquid chromatography with fluorimetric detection. J Chromatogr B 785: 219-226, 2003.

3. Beals, RK: Homocystinuria: a report of two cases and review of the literature. J Bone Joint Surg 51A: 1564-1572, 1969.

4. Bodamer OA, Sahoo T, Beaudet AL: Creatine metabolism in combined methylmalonic aciduria and homocystinuria. Ann Neurol 57: 557-560, 2005.

5. Brenton DP, Cusworth DC, Dent CE, Jones EE: Homocystinuria: clinical and dietary studies. Quart J Med. 35: 325-346, 1966.

6. Brenton DP, Dow CJ, James JIP et al: Homocystinuria and Marfan's syndrome: a comparison. J Bone Joint Surg. 54B: 277-298, 1972.

7. Bett EM: Homocystinuria with epilepsy. Proc Roy Soc Med 59: 484-486, 1966.

8. Burke JP, O'Keefe M, Bowell R, Naughten ER: Ocular complications of in homocystinuria.—early and late treated. Br J Opthalmol 73:427-431,1989.

9. Carson NAJ, Niell DW: Metabolic abnormalities detected in a survey of mentally backward individuals in Northern Ireland. Arch Dis Child 37: 505-513, 1962.

10. Chace DH, Hillman SL, Millington DS et al: Rapid diagnosis of homocystinuria and other hypermethionemias from newborn's blood spot by tandem mass spectrometry. Clin Chem 42: 349-355, 1996.

11. Chou S-M, Waisman HA: Spongy degeneration of the central nervous system: case of homocystinuria. Arch Pathol 79:357-363, 1965.

12. Collins JE, Benton DP: Pancreatitis and homocystinuria. J Inherit Metab Dis 13: 232-233, 1990.

13. Cross HE, Jensen AD: Ocular manifestations in the Marfan syndrome and homocystinuria. Am J Ophthalmol 75: 405-420, 1973.

14. Devlin AM, Hajipour L, Gholkar A et al: Cerebral edema associated with betaine treatment in classical homocystinuria. J Peidtar 14: 545-548, 2004.

15. Dunn HG, Perry TL, Dolman CL: Homocystinuria: a recently discovered cause of cause of mental defect and cerebrovascular thrombosis. Neurology 16:407-420, 1966.

16. Ekinci B, Apaydin H, Vural M, Ozekmekci S: Two siblings with homocystinuria presenting with dystonia and parkinsonism. Mov Disord 19: 962-964, 2004.

17. Febriani AD, Sakamoto A, Ono H et al: Determination of total homocysteine in dried blood spots using high performance liquid chromatography for homocystinuria newborn screening. Pediatr Int. 46: 5-9, 2004.

18. Gallagher PM, Naughten E, Hanson NQ et al: Characterizations of mutations in the cystathionine beta synthase gene in Irish patients with homocystinuria. Mol Genet Metab 65: 298-302, 1998.

19. Gerding H: Ocular complications and a new surgical approach to lens dislocation in homocystinuria due to cystathionine-beta-synthetase deficiency. Eur J Pediatr. 157 Suppl 2: S94-S101, 1998.

20. Gerritson T, Vaughn JG, Waisman HA: The identification of homocystine in the urine. Biochem Biophys, Res Commun 9: 493-496, 1962.

21. Gerritson T, Waisman HA: Homocystinuria, an error in the metabolism of methionine. Pediatrics 33: 413-420, 1964.

22. Gibson JB, Carson NAJ, Niell DW: Pathologic findings in homocystinuria. J Clin Pathol 17:427-437, 1964.

23. Godfrey M: The Marfan syndrome: In McKusick's Heritable Disorders of Connective Tissue. Beighton P, editor. St. Louis Mosby-Year Book, Inc. 1993, 51=122.

24. Harker LA, Slichter SJ, Scott CR, Ross R: Homocystinaemia: vascular injury and arterial thrombosis. N Engl J Med 291: 537-543, 1974.

25. Harrison DA, Mullaney PB, Mesfer SA et al: Management of ophthalmic complications of homocystinuria. Ophthalmology 105: 1886-1890, 1998.

26. Heil SG, Hogeveen M, Kluijtmans LA et al: Marfanoid features in a child with combined methylmalonic aciduria and homocystinura (CblC type) J Inherit Metab Dis 30: 811, 2007.

27. Hubmacher D, Tiedemann K, Bartels R et al: Modification of the structure and function of fibrillin-1 by homocysteine suggest a potential pathogenetic mechanism in homocystinuria. J Biol Chem 280: 34946-34955, 2005.

28. Isherwood DM: Editorial: Homocystinuria. Brit Med J: 313: 1025-1026, 1996.

29. Jensen AD, Cross HE: Surgical treatment of dislocated lenses in the Marfan syndrome and homocystinuria. Trans Am Acad Ophthalmol Otolaryngol 76: 1491-1499, 1972.

30. Kang AH, Trelstead RL: A collagen defect in homocystinuria. J Clin Invest 2571-2578, 1973.

31. Komrower GM, Wilson VK: Homocystinuria. Proc Roy Soc Med 56: 996-997, 1963.

32. Kraus JP, Janosik M, Kozich V et al: Cystationine beta-synthase mutations in homocystinuria. Hum Mutat 13: 362-75, 1999.

33. Laster L, Mudd SH, Finkelstein JD, Irreverre F: Homocystinuria due to cystathionine synthase deficiency: the metabolism of L-methionine. J Clin Invest 44: 1708-1719, 1965.

34. Lawson-Yuen A, Levy HL: The use of betaine in the treatment of elevated homocysteine. Mol Genet Metab 88:201-207, 2006.

35. Lee PJ, Briddon A: A rationale for cystine supplementation in severe homocystinuria. J Inherit Metab Dis. 30: 35-38, 2007.

36. Lee SJ, Lee DH, Yoo HW et al: Identification and functional analysis of cystathionine beta-synthase gene mutations in patients with homocystinuria. J Hum Genet. 50: 648-654, 2005.

37. Lerner-Ellis JP, Tirone JC, Pawelek PD et al: Identification of the gene responsible for methylmalonic aciduria and homocystinura cblC type. Nat Genet 38: 93-100, 2006.

38. Levy HL: Metabolic disorders in the center of genetic medicine. N Engl J Med 353:1968-1970, 2005.

39. Levy HL, Vargas JE, Waisbren SE et al: Reproductive fitness in maternal homocystinuria due to cystathionine beta-synthase deficiency. J Inherit Metab Dis 25: 299-314, 2002.

40. Li SC, Stewart pM: Homocystinuria and psychiatric disorder: a case report. Pathology 31: 221-224, 1999.

41. Lubec B, Fan-Kircher S, Lubec T et al: Evidence for McKusick's hypothesis of deficient collagen cross-linking in patients with homocystinuria. Biochem Biophys Acta. 1315: 159-162, 1996.

42. Majors AK, Sengupta S, Jacobsen DW, Pyeritz RE: Upregulation of smooth muscle cell collagen production by homocysteine-insight into the pathogenesis of homocystinuria. Mol Genet Metab 76: 92-99, 2002.

43. McKusick VA: Homocystinuria. In Heritable Disorders of Connective Tissue, 3rd Edition. St. Louis, Mosby, 1966, 150-178.

44. Miles EW, Kraus JP: Cystathionine beta-synthase: structure, function, regulation and locations of homocystinuria-causing mutations. J Biol Chem 279: 29871-29874, 2004.

45. Mudd SH: Vascularf diseae and homocysteine metabolism. New Engl J Med 313: 751-753, 1985.

46. Mudd SH, Finkelstein JD, Irreverre F, Laster L: Homocystinuria: an enzymatic defect. Science 143: 1443-1445, 1964.

47. Mudd SH, Levy HL, Abeles RH: A derangement in B12 metabolism leading to homcystinemia cysathioninemia and methlymalonic aciduria. Biochem Biophys Res Comm. 35: 121-126, 1969.

48. Mudd SH, Skovby F, Levy HL et al: The natural history of homocystinuria due to cystathionine β-synthase deficiency. Am J Hum Gent 37: 1-31, 1985.

49. Mulvihill A, Yap S, O'Keefe M et al: Ocular findings among patients with late-diagnosed or poorly controlled homcystinuria compared with a screened, well-controlled population. JAAPOS 5: 311-315, 2001.

50. Naughten ER, Yap S, Mayne PD: Newborn screening for homocystinuria: Irish and the world experience. Eur J Pediatr 157 Suppl 2: 584-587, 1998.

51. Peterschmitt MJ, Simmons JR, Levy HL: Reduction of false negative results in screening of newborns for homocystinuria. N Eng J Med 341: 1572-157, 1999.

52. Presley GD, Sidbury JB: Homocystinuria and ocular defects. Am J Opthalmol 63: 123-127, 1967.

53. Price J, Vickers CF, Brooker FK: A case of homocystinuria with noteworthy dermatological features. J ment Defic Res 12: 111-118, 1968.

54. Pyeritz, RE: Homocystinuria. In McKusick's Heritable Disorders of Connective Tissue. 5th Edition, Beighton, P Ed. St Louis, Mosby 1993,137-178.

55. Quere IC, Gris JC, Dauzat M: Homocystiene and venous thrombosis. N Engl J Med353: 1968-1970, 2005.

56. Refsum H, Fredriksen A, Meyer K et al: Birth prevalence of homocystinuria. J Pediatr 144: 830-832, 2004.

57. Ricci D, Pane M, Deodato F et al: Assessment of visual function in children with methylmalonic aciduria and homocystinuria. Neuropediatrics 36: 181-185, 2005.

58. Ryan MM, Sidhu RK, Alexander J, Megerian JT: Homocystinuria presenting as psychosis in an adolescent. J Child Neurol 17: 59-860, 2002.

59. Sakai Ly, Keene DR, Engvall E: Fibrillin, a new 350 kd glycoprotein is a component of extracelular microfibrils J Cell Biol 103, 2499-2509, 1986.

60. Schimke RN, McKusick VA, Pollack AD: Homocystinuria: a study of 38 cases in 20 families. JAMA 193: 711-719, 1965.

61. Schwahn BC, Hafner D, Hohlfeld T et al: Pharmacokinetics of oral betaine in healthy subjects and patients with homocystinuria. Br J Clin Pharmacol 55: 5-13, 2003.

62. Shelley WB, Rawnsley HM, Morrow G III: Pyridoxine-dependent hair pigmentation in association with homocystinuria. Arch Derm 106: 228-230, 1972.

63. Sinclair AJ, Barling L, Nightingale S: Recurrent dystonia in homocystinuria: a metabolic pathogenesis. Mov Disod 21: 1780-1782, 2006.

64. Shoji :YH, Takahashi T, Sato W et al: Acute life-threatening event with rhabdomyolysis after starting on high dose pyridoxine therapy in an infant with homocystinuria. J Inherit Metab Dis 21: 439-440,1998

65. Smith SW: Roentgen findings in homocystinuria. Am J Roentgenol 100: 147-154, 1967.

66. Tangerman A, Wilcken B, Levy HL et al: Methionine transamination in patients with homocystinuria due to cystathionine β-synthase deficiency. Metabolism 49: 1071-1077, 2000.

67. Topaloglu AK, Sansarick C, Snyderman SE: Influence of metabolic control on growth in homocystinuria due to cystathionine B-synthase deficiency. Pediatr Res 49: 796-798, 2001.

68. Streck EL, Delwing D, Tagliari et al: Brain energy metabolism is compromised by the metabolites accumulating in homocystinuria. Neurochem Int 43: 5970602, 2003.

69. Van Meurs JBJ, Dhonukshe-Rutten RAM, Pluijm SMF et al: Homocysteine levels and the risk of osteoporotic fracture. N Eng J Med 350: 2033-2041, 2004.

70. Walker BA, Beighton PH, Murdoch JL: The Marfanoid hypermobility syndrome. Ann Intern Med 71: 349-352, 1969

71. Wilcken DEL, Wilcken B: The natural history of vascular disease in homocystinuria and effects of treatment. J Inherited Metab Dis 20: 295-300, 1997.

72. Yap S: Classical homocystinuria: vascular risk and its prevention. J Inherit Metab Dis 26: 259-265, 2003.

73. Yap S, Boers GHJ, Wicken B et al: Vascular outcome in patients with homocystinuria due to cystathionine β synthase deficiency treated chronically: a multicenter observational study. Arterioscler Thromb Vasc Biol 21: 2080-2085, 2001.

74. Yap S, Naughten E: Homocystinuria due to cystathionine beta-synthase deficiency in Ireland: 25 years' experience of a newborn screened and treated population with reference to clinical outcome and biochemical control. J Inherit Metab Dis 21: 738-747, 1998.

75. Yap S, Rushe H, Howard PM, Naughten ER: The intellectual abilities of early-treated individuals with pyridoxine-non-responsive homocystinuria due to cystathionine beta-synthase deficiency. J Inherit Metab Dis 24: 437-447, 2001.

CHAPTER 4

Acromegaly: A Rare and Diverse Syndrome

Acromegaly was first defined in the mid 19th century and appears to be principally related to excessive growth hormone production. The condition is only rarely seen as genetic and is mostly related to a spontaneously developing pituitary adenoma. The disease occurs in two forms: a childhood disorder, which causes gigantism; and an adult form which is characterized by abnormal facies, calvarium, hands and feet and is associated with cardiac abnormalities and multiple foci of osteoarthritis. Over the past 30 years, the biochemical features of the abnormality have been clearly identified and drug therapy introduced to counter the abnormal effects of excessive growth hormone. Surgery was originally complex with a high rate of complications but in recent years minimally invasive trans-sphenoidal pituitary resection has been successful.

Nomenclature and history:

The term acromegaly comes from two Greek words: "akros" describing peripheral parts and "megaly" indicating enlargement or lengthening. The name was introduced by Pierre Marie in 1887 (57) and hence the disorder is known as Marie syndrome or in a somewhat different form, the Marie-Bamberger syndrome (4). Additional terms include gigantism, gigantosoma, giant hypersomia, somatomegaly, somatotropinoma, acromegalia syndrome, Carney complex and Erdheim syndrome (44,60).

A report by Toni et al (77) suggests that pictures of persons living in north and central America in pre-Colombian days strongly supported the existence of the diagnosis of acromegaly. Pierre Marie described and named the disorder in a seminal article published in 1886 and hence the eponymic term of "Marie's disease" (57). It should be noted however that the unique characteristics of the disorder were recognized by Verga in 1864 (82) and by Brigidi in 1881 (15). In 1890 (56) Marie defined the pathology and in 1891, he and his colleague Marinesco (58) clearly identified the syndrome, but did not relate it to the pituitary gland. It was Massalongo in 1892 (59), who first identified the cause as related to excessive growth hormone and shortly thereafter Benda (7) described the pathologic histology of the pituitary gland. In 1910, Glick and co-workers (39) were able to perform an immunoassay of human growth hormone. In 1921, Evans and Long (35) showed that growth markedly increased in the rat with treatment with pituitary growth hormone.

Roth et al in 1967 (72) clearly described the increase in growth hormone production in patients with acromegaly. In the early part of the 20th century, Harvey Cushing (29) and several other surgeons (21,41) suggested that surgical removal of the pituitary gland could cure the patients and the trans-sphenoidal minimal approach could be a successful method of eliminating the disease (42). Radiation therapy was introduced shortly thereafter and became a standard approach for many of the patients but the complications were troublesome (44,51,55,74). In 1973, Hall and Besser (40) and the following year, Besser and colleagues (9) reported that parenteral somatostatin infusions could reduce growth hormone levels; and soon thereafter somatostatin analogue therapy was introduced (5,11,44,48,60).

Other syndromes which were noted to be associated with acromegalic findings included those reported by Erdheim in 1931 (31), Touraine, Solente and Gole in 1935 (79), Carney in 1979 (18) and Bhansali et al (8) for patients with McCune-Albright syndrome.

Biochemical causation:

Despite diverse histologic abnormalities, excessive production of growth hormone (GH) is the factor which causes both childhood gigantism and late in adulthood acromegaly (2,60). Normally the hypothalamus releases a material known as growth hormone releasing hormone (GHRH), which causes the pituitary gland to increase the level of GH production (60). In turn, GH activates elevated levels of liver insulin-growth factor-I (IGF-I) otherwise known as somatomedin C, which is responsible for growth of organs including bone (17,44,60,84). The IGF-I appears to be the key factor in the development of the syndrome of acromegaly (60,84). Levels of IGF-I are consistently elevated in the disease and hence are used to monitor treatment success (17,23). Another hormone produced by the hypothalamus is somatostatin, which inhibits GH production and release (48,60). Normally the levels of GH, GHRH, IGF-I and somatostatin are regulated by each other and by other body activities including sleep, exercise, food intake and blood sugar levels (44,60,84).

The causes of elevated growth hormone include the following:

1. Primary pituitary tumors particularly adenomas which consist of somatotrophs (growth hormone secreting cells) or mammosomatotrophs (growth hormone and prolactin secreting cells) (2,33,44,60,73).
2. Familial disease which occurs mostly in children is probably a somatotropinoma and is possibly a gene error on 11q13 or 13q14 (1,37,45,63,76).
3. Some patients with the Carney complex consisting of myxomas, pigmentation, gastric lesions and extra-adrenal paraganglioma, may also have acromegalic features (13,18,19).
4. Several patients with McCune-Albright syndrome have been found to have increased GH and findings consistent with acromegaly (8,16,38). The findings in female patients with this disorder include café au lait spots, often severe fibrous dysplasia and premature puberty.
5. Erdheim syndrome, originally described by Jakob Erdheim in 1931 consists of acromegaly with bone and cartilage hypetrophy of the clavicle, vertebral bodies and intervertrebal disks (31).

6. Touraine-Solente-Gole syndrome originally described by the three authors in 1935, is an autosomal dominant condition which consists of clubbing of the digits, long extremities, periosteal new bone formation, coarsening of the facial features with furrowing of the face, forehead and scalp (79).

7. Some disorders of other organs, mostly neoplastic that produce increased amounts of growth hormone (10,62).

Clinical presentations for acromegaly:

There are two principal presentations for acromegaly: gigantism in children, and acromegalic changes in adults with late onset disease. In addition as defined in the previous section, entities such as the Erdheim disease, Carney complex, McCune-Albright syndrome and Touraine-Solente-Gole disorder all have arrays of abnormalities which include aspects of acromegaly.

Gigantism defines the abnormally high linear growth in children due to the excessive action of IGF-I based on increased amounts of GH or GHRH, which may occur for a variety of reasons (1,2,5,26,44,52,60). Most often the gigantism is caused by GH-secreting pituitary adenoma in a growing child. In addition, however children may rarely be affected by McCune-Albright syndrome, multiple endocrine neoplasia type I (MEN-I), neurofibromatosis or Carney complex (18,19,44,60,80). Gigantism is exceedingly rare and is less frequently encountered than adult acromegaly. There is no racial predilection and men and women are affected equally. The disorder may begin at any age prior to epiphyseal fusion. Characteristically the children grow rapidly and display tall stature, often well over 6 ½ feet tall. Mild to moderate obesity is common as is soft tissue hypertrophy and exaggerated growth of hands and feet. Cardiac and joint problems are uncommon in this group although they frequently report back pain, which is sometimes disabling (12,60). The bones are normal in appearance on X-ray other than being larger although the ribs and spine are sometimes increased in size and density (52,60,83). A number of famous people showed changes supportive of the diagnosis of childhood onset acromegaly. These included: actor Richard Kiel (7'2" tall); actor Matthew McGrory (7'6" tall); NBA basketball player Gheroghe Muresan (7'7" tall); basketball player Sun Ming Ming (7'9" tall); and composer Sergei Rachmaninoff. Acromegaly has been described in dogs (30).

Adult acromegaly is more common and at times much more severe than pituitary gigantism and occurs equally in males or females. The disorder may be present for years prior to discovery which often occurs in middle age (5). The patients develop sponginess and puffiness of hands and feet and this is particularly evident in the heel pads (36). Increased body hair is often noted. Colon polyps are common (25,34) and occasional colon cancer has been described (43). The patients often complain of increased sweating and oily skin. The finger and toe-nails become thickened. Facial features are often noticeably changed with coarsening, large pores, swollen eyelids, enlargement of nose and changes in the voice (22,49,64,75,84). The patients complain of fatigue, back and joint pain, increased space between teeth, polyuria, polydipsia, decreased libido and impotence, sleep apnea, depression and muscle weakness (26,27,84). Osteoarthritic changes in the knee, hip, shoulder, spine and feet and are often disabling (12,14). Small sessile and pedunculated skin lesions may be present along with hypertrichosis (22). Cardiac abnormalities are common and frequently life threatening (28,84). They consist of mitral valve regurgitation (68,80) and cardiomyopathy (84), which may cause an early demise (85). Increased numbers of breast and colon cancers are described and occasional cases of osteosarcoma are discovered in an affected bone (6,43,53). If the pituitary adenoma continues to grow, neurologic symptoms including mental aberrations, hearing and sight disturbances and emotional changes may cause considerable distress (84).

Although the other diseases that cause acromegaly are very rare, review of family history will help to define the genetic forms. Patients who have Carney complex may have a pulmonary chondroma, gastric leiomyosarcoma and extra-adrenal paraganglioma (19). Female patients with McCune-Albright syndrome may be identified by a history of precocious puberty, café au lait spots and radiographic changes consistent with fibrous dysplasia (16). Patients with the Erdheim syndrome often have bone and cartilage hypertrophic changes (31).

Laboratory studies will almost always show increased IGF-I and often increased GH (5,17,23,33,39,60,72). Imaging studies may show osteophyte formation at the articular margins of many joints and enlargement of the tufts of the distal phalanges (83). Studies of the hands and particularly the feet may show increased thickness of the soft tissues overlying the bones (12,14,36). The ribs are often enlarged particularly at the sites of the costo-chondral junctions (83). Scoliosis and abnormal vertebral structure are commonly present (14). Studies of the skull may show damage to the sella turcica caused by enlargement of the pituitary adenoma and sometimes marked enlargement of the mandible (49). Malformation of the teeth is a common finding (60,84).

Treatment of patients with acromegaly:

Over the many years of observation of the disease there are basically three forms of treatment, which seem to be effective in the management of patients with these entitities.

Radiation of the pituitary adenoma was introduced early and remains somewhat helpful in management (44,51,55,74). Originally radiation was given by standard external beam but the complications in terms of skin, hair and bone changes were troublesome (44,55). In recent years, the gamma knife technique has been introduced and appears to be more effective (20).

Surgery was first introduced at the beginning of the 20th century by a number of surgeons including Harvey Cushing and consisted initially as trans-craniotomy (21,29,38,41,74). It became apparent that the complications of this surgery were a problem so that the trans-sphenoidal approach became the choice and is still used by neurosurgeons today, with a high rate of success and low complication rate (32,42,47,71).

Chemotherapy consists almost entirely of agents described as somatostatin analogues. These include octoreotide, pegvisomant, lanreotide and bromcriptine, all of which decrease growth hormone production (3,11,24,38,46,48,50,54,61,65-67,69,78,81). The side effects of these agents are sometimes troublesome, but in general they reduce cardiac complications of the disease and improve the patient's emotional and physical status. They may be utilized with radiation or following surgery if the procedures do not decrease the IGF-I concentration (24,38,60,67).

Regardless of treatment, patients must be carefully followed and particularly repeatedly studied for cardiac abnormalities (60), gastrointestinal and other tumor development (53,70) and bone and joint disorders (12,14,26,52).

Conclusions:

Most physicians and especially orthopaedic surgeons have little contact with patients with acromegaly. The disorder is rare and the diagnosis is difficult to make since many of the patients with the problem have little evident difficulty. The adult form of the disease which is the most common, usually occurs somewhat late in life and has a series of symptoms and findings that might be considered both by the patient and physician as part of "growing old". The subtlety of the findings, which are not greatly different from aging, make it hard to identify the disorder. Use of examination of the serum for growth hormone or even more logically IGF-I, could be helpful in establishing the diagnosis.

Once the diagnosis is discovered, it is essential to obtain imaging studies of the pituitary gland and if possible plan for removal, probably best performed by the trans-sphenoidal approach. Gamma knife radiation may be the best approach for older individuals. The use of one of the somatostatin analogues is very helpful in the care of these patients but may not completely remove the primary lesion and may over time have additional complications. The key issue perhaps is to define and hopefully completely eliminate the disease before colon cancer or cardiac abnormalities destroy the patient. That's the challenge for this rare and really quite bizarre clinical entity.

References:

1. Abbassioun K, Fatourehchi V, Amirjamshidi A, Meibodi NA: Familial acromegaly with pituitary adenoma: report of three affected siblings. J Neurosurg 64: 510-512, 1986.
2. Asa SL, Ezzat S: Genetics and proteomics of pituitary tumors. Endocrine 28: 43-7, 2005.
3. Attanasio R, Bardelli R, Pivonello R et al: Lanreotide 60 mg, a new long acting formulation: effectiveness in the chronic treatment of acromegaly. J Clin Endocriinol Metab 88:5258-5265, 2003.
4. Bamberger E: Über knochenveranderungen gei chronischen lungen und herzkrankheiten Zachr Klin Med 18: 193-217, 1891.
5. Barkan AL: Acromegaly: diagnosis and therapy. Endocrinol Metab Clin North Am 18: 277-310, 1989.
6. Barzilay J, Heatley GJ, Cushing GW: Benign and malignant tumors in patients with acromegaly. Arch Intern Med. 151: 1629-1632, 1991.
7. Benda C: Beitrage zur normalen und pathologischen histology der menschlichen hypophysis cerebri. Klin Wochenschr 36: 1205, 1900.
8. Bhansali A, Sharma BS, Sreenivasulu P et al: Acromegaly with fibrous dysplasia: McCune-Albright syndrome. Endocrin J 50: 793-799, 2003.
9. Besser GM, Mortimer CH, Carr D et al: Growth hormone release inhibiting hormone in acromegaly. Br Med J 1: 352-355, 1974.

10. Beuschlein F, Stasburger CJ, Siegerstetter V et al: Acromegaly caused by secretion of growth hormone by a non-Hodgkin's lymphoma. N Engl J Med 342: 1871-1876, 2000.

11. Bevan JS: The antitumoral effects of somatostatin analog therapy in acromegaly. J Clin Endocrinol Metab. 90: 1856-1863, 2005.

12. Bluestone R, Bywaters EGL, Hartog M et al. Acromegalic arthropathy. Ann Rheum Dis 30: 243-258, 1971.

13. Boikos SA Stratakis CA: Carney complex: the first 20 years. Curr Opin Oncol 19: 24-29, 2007.

14. Bonadonna S, Mazziotti G, Nuzzo M et al: Increased prevalence of radiological spinal deformities in active acromegaly: a cross-sectional study in postmenopausal women. J Bone Miner Res 20: 1837-1844, 2005.

15. Brigidi V: Studii anatomo-patologici sopra un uomo divenuto stranamente deforme per cronica infermita. Arch Scuola Anat Patol Firenze 65-92, 1881.

16. Brockmann H, Joe A, Palmedo H, Biersack H-J: A patient with acromegaly presenting with polyostotic fibrous dysplasia on bone scan: McCune-Albright syndrome. Clin Nucl Med. 30: 813-815, 2005

17. Brooke AM, Drake WM: Serum IGF-I levels in the diagnosis and monitoring of acromegaly. Pituitary 10: 173-179, 2007.

18. Carney JA: The triad of gastric epitheliod leiomyosarcoma, functioning extra-adrenal paraganglioma and pulmonary chondroma. Cancer 43: 374-382, 1979.

19. Carney JA: The Carney complex (myxomas, spotty pigmentation, endocrine overactivity ands schwannomas) Dermatol Clin 13: 19-26, 1995.

20. Castinetti F, Taieb D, Kuhn JM et al: Outcome of gamma knife radiosurgery in 82 patients with acromegaly: correlation with initial hypersecretion. J Clin Endcrinol Metab. 90: 4483-4488, 2005.

21. Caton R, Paul FT: Notes on a case of acromegaly treated by operation. Br Med J 2: 1421-1423, 1893.

22. Centurion SA, Schwartz RA: Cutaneous signs of acromegaly. J Dermatol 41: 631-634, 2002.

23. Clemmons DR: Quantitative measurement of IGF-I and its use in diagnosing and monitoring treatment of disorders of growth hormone secretion. Endocr Dev 9: 55-65, 2005.

24. Colao A, Attanasio R, Pivonello R: Partial surgical removal of growth hormone—secreting pituitary tumors enhances the response to somatostatin analogs in acromegaly. J Clin Endocrinol Metab 91: 85-92, 2006.

25. Colao A, Balzaono A, Ferone D et al: Increased prevalence of colonic polyps and altered lymphocyte subset pattern in the colonic lamina propria in acromegaly. Clin Endocrinol (Oxf) 47: 23-28, 1997.

26. Colao A, Ferone D, Marzullo P, Lombardi G: Systemic complications of acromegaly: epidemiology, pathogenesis and management. Endocrin Rev 25: 102-152, 2004.

27, Colao A, Pivonello R, Marzullo P et al: Severe systemic complications of acromegaly. Endocrinol Invest. 28 (5 Suppl): 65-77, 2005.

28. Colao A, Spinelli L, Cuocolo A et al: Cardiovascular consequences of early-onset growth hormone excess. J Clin Endocr Metab 87: 3097-3104, 2002.

29. Cushing H: Partial hypophysectomy for acromegaly: with remarks on the function of the hypophysis. Ann Surg 50: 1002-1017, 1909.

30. Eigenmann JE: Acromegaly in the dog. Vet Clin North Am Small Anim Pract 14: 827-836, 1984.

31. Erdheim J: Über Wirbersaulenveranderungen bei akromegalie. 281: 197-206, 1931.

32. Erturk E, Tuncel E, Kiyici S et al: Outcome of surgery for acromegaly performed by different surgeons. Pituitary 8: 93-97, 2005.

33. Ezzat S, Serri O, Chik CL et al: Canadian consensus guidelines for the diagnosis and management of acromegaly. Clin Invest Med 29: 29-39, 2006.

34. Ezzat S, Strom C, Melmed S: Colon polyps in acromegaly. Ann Intern Med 114: 754-755. 1991.

35. Evans HM, Long JA: The effect of the anterior lobe of the pituitary administered intra-peritoneally upon growth, maturity and oestrus cycle of the rat. Anat Rec 21: 62-70, 1921.

36. Fields ML, Greenberg BH, Burkett LL: Roentgeonographic measurement of heel-pad thickness in the diagnosis of acromegaly. Am J Med Sci 254: 528-533, 1967.

37. Gadelha MR, Une KN, Rohde K, et al: Isolated familial somatotropinomas: establishment of linkage to chromosome 11g13.1-11q13.3 and evidence for a potential second locus at chromosome 2p16-12. J Clin Endocr Metab. 85: 707-714, 2000.

38. Galland F, Kamenicky P, Affres H et al: McCune-Albright syndrome and acromegaly: Effects of hypothalmopituitary radiotherapy and /or pegvisomant in somatostatin analog-resistant patients. J Clin Endocrin Metab 91: 4957-4961, 2006.

39. Glick SM, Roth J, Yalow RS, Berson SA: Immunoassay of human growth hormone in plasma. JAMA 5:772-774, 1910.

40. Hall R, Besser GM Schally AV, Coy DH et al: Action of growth hormone-release inhibitory hormone in healthy men and in acromegaly. Lancet 2: 581-584, 1973.

41. Halstead AE: Remarks on the operative treatment of tumors of the hypophysis. With a report of two cases operated on by an oronasal method. Trans Am Surg Assoc 28: 73-93, 1910.

42. Hirsch O: Endonasal method of removal of hypophyseal tumors. JAMA 5: 772-774, 1910.

43. Ituarte EA, Petrini J, Hershman JM: Acromegaly and colon cancer. Ann Intern Med 101: 627-628, 1984.

44. Jane JA Jr., Thapar K, Laws ER Jr. Acromegaly: historical perspectives and current therapy. J Neurooncol 54: 129-137, 2001.

45. Jones MK, Evans PJ, Jones IR, Thomas JP: Familial acromegaly. Clin Endocr. 20: 355-358, 1984.

46. Jorgensen JO, Feldt-Rasmussen U, Frystyk J et al: Co-treatment of acromegaly with a somatostatin analog and growth hormone receptor antagonist. J Clin Endocrinol Metab. 90: 5627-5631, 2005.

47. Klibanski A, Zervas NT Diagnosis and managment of hormone—secreting pituitary adenomas. N Engl J Med 324: 822-831, 1991.

48. Kopchick JJ, Parkinson C, Stevens EC, Trainer PJ: Growth hormone receptor antagonists: discovery, development and use in patients with acromegaly. 23: 623-646, 2002.

49. Kunzler A, Farmand M: Typical changes in the viscerocranium in acromegaly. J Craniomaxillofac Surg 19: 332-340, 1991.

50. Lamberts SW, van der Lely AJ, de Herder WW et al: Octreotide. N Engl J Med 334: 246-254, 1996.

51. Lamberg BA, Kivikangas V, Vartianen et al: Conventional pituitary irradiation in acromegaly. Effect on growth hormone and TSH secretion. Acta Endocrinol (Copenh) 82: 267-281, 1976.

52. Lieberman SA, Bjorkengren AG, Hoffman AR: Rheumatologic and skeletal changes in acromegaly. Endocrinol Metab Clin North Am. 21: 615-631, 1992.

53. Lima GA, Gomes EM, Nunes C et al: Osteosarcoma and acromegaly: a case report and review of the literature. J Endocrinol Invest 29: 1006-1011, 2006.

54. Lindberg-Larsen R, Moller N, Schmitz O et al: The impact of pegvisomant treatment on substrate metabolism and insulin sensitivity in patients with acromegaly. J Clin Endocrinol Metab 92: 1724-1728, 2007.

55. Macleod AF, Clarke DG, Pambakian H et al: Treatment of acromegaly by external irradiation. Clin Endocrin (Oxf).30: 303-314, 1989

56. Marie P: De l'osteo-arthropathie hypertrophiante pneumique Rev Med Paris 10: 1-36, 1890.

57. Marie P: Sur deux cas d'acromegalie: hypertrophie singuliere non congenitale des extremites superieurs, inferieurs et cephalique Rev Med Paris 6:297-333, 1886.

58. Marie P, Marinesco G: Sur l'anatomie pathologique de l'acromegalie. Arch Med Exp Anat 3: 539-565, 1891.

59. Massalongo R: Sull' acromegalia. Riforma Med 8: 74, 1892.

60. Melmed S: Medical Progress: Acromegaly. N Engl J Med 355: 2558-2573, 2006.

61. Melmed S, Casnueva F, Cavagnini F et al: Consensus statement: medical management of acromegaly. Eur J Endcrinol 153: 737-740, 2005.

62. Melmed S, Ezin C, Kovas K et al: Acromegaly due to secretion of growth hormone by an ectopic pancreatic islet-cell tumor. N Engl J Med 312: 9-17, 1985.

63. McCarthy MI, Noonan K, Wass JAH, Monson JP: Familial acromegaly: studies in three families. Clin Endocrin 32: 719-728, 1990

64. Motta S, Ferone D, Colao A et al: Fixity of vocal cords and laryngocele in acromegaly. J Endocrinol Invest. 20: 672-674, 1997.

65. Newman CB, Melmed S, George A et al: Octreotide as primary therapy for acromegaly. J Clin Endocrinol Metab. 20: 672-674, 1997.

66. Page MD, Milward ME, Taylor A et al Long-term treatment of acromegaly with a long-acting analogue of somatostatin. Octreotide. Quart J Med 74: 189-201, 1990.

67. Parkinson C, Burman P, Messig M, Trainer PJ: Gender, body weight, disease activity and previous radiotherapy influence the response to pegvisomant. J Clin endocrine Metab. 92: 190-195, 2007.

68. Pereira AM. Van Thiel SW, Lindner JR er al: Increased prevalence of regurgitant valvular heart disease in acromegaly. J Clin Endocrinol 89: 71-75, 2004.

69. Pereira JL, Rodriguez-Puras MJ, Leal-Cerro A et al: Acromegalic cardiopathy improves after treatment with increasing doses of octreotide. J Endocrinol Invest. 14: 17-23, 1991.

70. Perry JK, Emerald BS, mertani HC, Lobie PE: The oncogenic potential of growth hormone. Growth Horm IGF Res. 16: 277-289, 2006.

71. Ross DA, Wilson CB: Results of transsphenoidal microsurgery for growth hormone—secreting pituitary adenoma in a series of 214 patients. J Neurosurg 68: 854-867, 1988.

72. Roth J, Glick SM, Hollander CS: Acromegaly and other disorders of growth hormone secretion. Ann Intern Med 66: 760-788, 1967.

73. Sata A, Ho KK: Growth hormone measurements in the diagnosis and monitoring of acromegaly. Pituitary 10: 165-172, 2007.

74. Sheline GE, Goldberg MB, Feldman R: Pituitary irradiation for acromegaly. Radiology 76: 70-75, 1961.

75. Sneppen SB, Main KM, Juul A et al: Sweat secretion rates in growth hormone disorders. Clin Endocrinol (Oxf). 53: 601-608, 2000.

76. Soares BS, Eguchi K, Frohman LA: Tumor deletion mapping on chromosome 11q13 in eight families with isolated familial somatotropinoma and 15 sporadic somatotropinomas. J Clin Endocr Metab 90: 6580-6587, 2005.

77. Toni R, Ghigo E, Roti E, Lechan RM: Endocrinology and art. Acromegaly and goiter in the Pre-Colombian, Mesoamerican population. m J Endocrinol Invest. 30: 169-170, 2007.

78. Trainer PJ, Drake WM, Katznelson L et al: Treatment of acromegaly with the growth hormone-receptor antagonist pegvisomant. N Engl J Med 342: 1171-1177, 2000.

79. Touraine A, Solente G, Gole L: Un syndrome ostodermpathique. La pachydermie plicaturee avec pachyperiostose des extremities. Presse Med 42: 1820-1824, 1935.

80. Van der Klaauw AA, Bax JJ, Roelfsema F et al: Uncontrolled acromegaly is associated with progressive mitral valvular regurgitation. Growth Horm IGH Res 16: 101-107, 2006.

81. Vance ML, Harris AG: Long-term treatment of 189 acromegalic patients with somatostatin analog octreotide. Results of the International Multicenter Acromegaly Study Group. Arch Intern Med 151: 153-1178, 1991.

82. Verga A: Caso singolare diprospectasia. Rend R 1st Lombardo Classe Sc Mat Nat 1: 111-117, 1864.

83. Woo CC: Radiological features and diagnosis of acromegaly. J Manipulative Physiol Ther 11: 2306-213, 1988.

84. Woodhouse LJ, Makherjee A, Shalet SM, Ezzat S: The influence of growth hormone status on physical impairments, functional limitations and health-related quality of life in adults. Endocrin Rev 27: 287-317, 2006.

85. Wright AD, Hill DM, Lowy C, Fraser TR: Mortality in acromegaly. Q J Med 39: 1-16. 1970.

CHAPTER 5

Arthrogryposis Multiplex Congenita

Introduction:

Arthrogryposis multiplex congenita was first described by Adolph Wilhelm Otto in 1841 in an article written in Latin and strangely titled "A human monster with inwardly curved extremities". It is ordinarily defined as a rare entity of unknown origin usually identified in newborn infants and characterized by multiple joint contractures and poorly developed muscles, which appears to result from neurogenic or myopathic alterations. The disease has a vast array of etiologic causes, some of which are genetic and some which include oral, facial, cardiac, gastrointestinal or pulmonary disorders in addition to the muscular defects. Dislocations of joints are common and deformities of the hands and feet are often striking. The children generally have a normal intelligence but a large percentage of them die of disease early in the course. There are currently no approaches to treatment other than pediatric, orthopaedic or physiotherapeutic management programs, which sometimes restore the skeletal structure but do not have much effect on muscular function.

Nomenclature and history for arthrogryposis multiplex congenita

A large number of terms have been defined for arthrogryposis multiplex congenita based in part on the vast array of entities, which seem to ultimately result in the characteristic extremity abnormalities. These include Otto syndrome, Guerin-Stern syndrome, Pena-Shakier syndrome, Rocher-Sheldon syndrome, amyoplasia, arthromyoplasia congenita, myodysplasia, neuro-arthrodysplasia, pterygoarthromyodysplasia, mylophagism, dysplasia-myo-osteo-articularis and myodystrophia fetalis deformans.

As indicated in the Introduction, the disease entity was first described by Adolph Wilhelm Otto in 1841 (47), who described the patient in his report as "a human monster with inwardly curved extremities". Guerin in 1822 further described the monsters and added the concept of a congenital abnormality. Schanz in 1897 (55) described another case and Rocher in 1913 (52) termed the process as "multiple congenital articular rigidities". Stern in 1923 (62) proposed the name "arthrogryposis multiplex congenita". The term arthrogryposis is a Greek expression defining the disorder as "hooked or curved joints". Lewin in 1925 (37) reviewed the orthopaedic characteristics of the disease. Sheldon in 1934 (57) is credited with the first clear description of a child with the disorder but he also introduced the term amyoplasia, based on his belief that the disease is muscular in origin, rather than neurologic. Middleton in 1934 (42) studied muscle tissues from patients and defined the disease as of muscular origin. Over the next 20 years, Brandt in 1947 (14), Adams et al in 1953 (1) and Wolf and coworkers in 1955 (68) attempted to define a neurologic or myogenic cause for the entity.

Genetic features for arthrogryposis multiplex congenita

The term arthrogryposis multiplex congenita includes a heterogeneous array of disorders, all of which have as their presenting finding congenital joint contractures, principally affecting the hands, feet, calves, arms and to lesser extent the upper parts of the limbs, pelvis and spine. In the past 20 years, the entity has been defined as occurring in association with over 250 specific disorders with arthrogrypotic features, some of which are genetic in origin (5,6,7,11,12,13,16-18,27,49,53,64). It was Judith Hall (27,18) and Michael Bamshad (5) however, who first proposed that there are three causes for the disorder: Type I-myopathic (limbs alone, also known as amyoplasia); Type II-myopathic but including lesions in other body areas; and Type III-myopathic but also primarily neuropathic.

The disease is common in some countries, which have an isolated population especially including Finland (48) and the Bedouin community in Israel (31,58,65). For most of the patients the disease is equally divided by gender but there are some X-linked genetic errors, which make it more frequently present in males (27,28). Intelligence is not usually impaired. Of some importance is the fact that Type III disease has a poor survival rate with over 50% of the patients dying within a year after birth (7,16,27,33,51). Many of the cases have no genetic characteristics and in fact may be related to maternal disorders including myasthenia gravis, multiple sclerosis, increased uterine pressure, hydramnios, infections or fevers (9,30,38,50). Some patients with arthrogrypotic changes particularly in the extremities have other disorders including Antley Bixler syndrome, Conradi-Hunermann disease, Gordon syndrome, Kniest syndrome, Holt-Aram disease, Beals syndrome, Turner syndrome, Moebius syndrome Prader-Willi abnormalities, Sturge-Weber disease, Weaver syndrome and Zellweger syndrome (5,7,8,16-18,26-28,33,34,44,64,66). Based on other features, these disorders may be recognized as distinct from arthrogryposis multiplex congenita and are basically excluded from discussion in this chapter. It is important however to attempt to rule them out when patients are evaluated. It should also be noted that a similar disease has been reported to occur in some animal species as well (40,64).

Four genetic loci associated with autosomal recessive transmission have been identified (6,27,28,33,34,44,58,64,65): lethal congenital contracture syndrome type 1 (chromosome 9q34); lethal congenital contracture syndrome type 2 (chromosome 12q13); the Pena-Shokeir phenotype neurogenic type (chromosome 5q35) (26); and arthrogryposis-renal dysfunction-cholestasis syndrome (chromosome 15q26.1). In addition however at least 35 other genetic abnormalities have been identified in patients with the disorders but at this point there is still limited information as to relationship to causation or clinical presentation of the various entities. Clearly, to quote a statement from an article written by Wesley Bevan, Judith Hall and Michael Bamshad (13) and published in the Journal of Pediatric Orthopaedics in 2007, "the molecular basis of most genetic arthrogryposes is an active area of investigation".

Clinical presentation of patients with arthrogryposis multiplex congenita:

Infants with Type I arthrogryposis multiplex congenita (amyoplasia) are most often strikingly abnormal in limb structure at birth (5.7,11,12,16,19,20,27,35,39,54,56,66). The children usually have average intelligence, are somewhat smaller in height and weight than normal infants and have sometimes markedly decreased muscle mass. Most often they have truncal sparing with most of the problems in the limbs (13,19). The arms are symmetrically affected and display down-sloping shoulders and internally rotated upper extremities (16,19). Elbows are normally extended and forearms pronated with marked flexion deformities of the wrists and hands (10,41,67). Lower limb involvement shows either flexed or extended knees, flexed and externally rotated hips and severe internal rotation and clubfoot deformities of the feet (23,39,43,69). The hips may be subluxated or dislocated (59,60,69). The trunk is generally spared.

Children with Type II disease, in addition to the motor problems and limb deformities seen in those with Type I disease, are most often short in stature and present with craniofacial deformities (24,29,36,49,61). The changes consist of flat nasal bridges, facial hemangiomas and jaw deformities, cleft palate, dental abnormalities, and small eyes with corneal opacities. Respiratory problems include tracheal and laryngeal stenosis and a weak diaphragm. Renal problems may occur and increase the risk of poor survival (16,17). Some of these infants, especially those who are classified as Pena-Shokeir disease, have a high death rate in association with cardiac abnormalities (26,46,49,51,53).

Children with Type III neuropathic disease have a high early death rate (over 50%). In addition to the amyoplastic abnormalities in the limbs they have loss of vigor, lethargy, absent reflexes, sensory deficits and sometimes seizures or partial paralysis (5,7,14,27, 51,53,68).

Imaging studies are sometimes quite striking related to the remarkably deficient muscular structure, the subluxation of the hips and deformities of the knees (13,19). The changes in the hands and especially the feet are quite dramatic (4,10,11,13,16,19,41,67). The feet show extraordinary deformities consistent with severe talipes equinovarus (23,27,39,45).

Electromyographic studies show very poor muscle strength and in some cases, particularly for children with Type 3 disease, very poor neural function (32,63). Muscle biopsies may show an irregular derangement of a diminished number of muscle fibers and some excess accumulation of fat (1,7,8,32,51,63).

Treatment of patients with arthrogryposis multiplex congenita:

Currently there are no drugs or genetic alteration therapies available for patients with this disorder. If the children have severe respiratory or cardiac abnormalities some forms of supportive treatment may be helpful, but most of the treatment protocols available are for amyoplastic disease, chiefly affecting the hands, feet, hips, knees and elbows. An exercise protocol is sometimes helpful and bracing may make it possible for these unfortunate children to use their hands, stand and sometimes to walk (25,59,60,66,69).

Orthopaedic operative procedures have been quite helpful in restoring and stabilizing the defective anatomical parts. Specifically, surgical corrections of the clubfeet have been performed and several authors have indicated that talectomies and arthrodeses have allowed the patients to ambulate (13, 15,21,23,45). Hand and wrist surgery has been helpful in restoring function, chiefly by the use of muscle transplants and ligamentous rearrangements (4,10,12,41). Shoulder surgery has also been helpful (4,67). Of considerable importance are hip reconstructions for dislocations (2,3,59,60,69) and realignment procedures for knee abnormalities (43,66). These operations have often allowed the patients to stand and ambulate.

It should be also noted that anesthesia is somewhat risky for these patients, especially for those with Type II or Type III disease (2,310,25,36,59,66). Respiratory and cardiac problems are very significant issues for these patients and should be carefully evaluated prior to surgery.

Conclusions:

Arthrogryposis multiplex congenita is a mysterious and complex disease of infants with very limited information about causation, presentation and outcome. It appears to occur in three major forms: Type I myopathic (limbs alone, also known as amyoplasia); Type II myopathic but associated with lesions in other body sites; Type III myopathic but associated with neuropathy. The Type I disease is often not fatal, but the other two are and especially the Type III. Despite multiple attempts to further define the disorder, the causation is still very confusing. There are suggestions regarding gene errors, but maternal health issues appear to have an effect as well. What is most striking is the fact that the myopathic presentation principally affecting the limbs and especially, the hands and feet may also occur with a large number of other clinical disorders. The relationship between these entities and the syndrome of arthrogryposis multiplex congenita, first identified in the early 18th century, cannot as yet be solved. Perhaps even more disturbing is that with the exception of orthopaedic corrective surgery for the extremities there is no identifiable treatment for these terribly deformed and disabled children. Quite clearly we need to assess how best to approach the children with treatments that might allow them to live and equally importantly, live well. It is certainly a great challenge for the pediatricians, orthopaedists and neurologists who have to care for the afflicted children and their often very distressed families.

References:

1. Adams RD, Denny-Brown D, Pearson CM: Diseases of Muscle. A Study in Pathology. New York, Paul B. Hoeber, Inc 1953: 229-235.
2. Akazawa H, Oda K, Mitani S et al: Surgical management of hip dislocation in children with arthrogryposis multiplex congenita. J Bone Joint Surg 80B: 636-640, 1998.

3. Asif S, Umer M, Beg R, Umar M: Operative treatment of bilateral hip dislocation in children with arthrogryposis multiplex congenita. J Orthop Surg 12: 4-9, 2004.

4. Axt MW, Niethard FU, Doderlein L, Weber M: Principles of treatment of the upper extremity in arthrogryposis multiplex congenita type 1. J Pediatr Orthop B 6: 179-185, 1997.

5. Bamshad M, Jorde B, Carey JC: A revised and extended classification of the distal arthrogryposes. Am J Med Genet 65: 277-281, 1996.

6. Bamshad M, Watkish WS, Zengerf RK et al: A gene for distal arthrogryposis type I maps to the percentromeric region of chromosome 9. Am J Hum Genet 55: 1153-1158, 1994.

7. Banker BQ: Neuropathologic aspects of arthrogryposis multiplex congenita Clin Orthop 194: 30-43, 1985.

8. Banker BQ: Arthrogryposis multiplex congenita: spectrum of pathologic changes. Hum Pathol 17: 656-672, 1986.

9. Barnes PR, Kanabar DJ, Brueton L et al: Recurrent congenital arthrogryposis leading to a diagnosis of myasthenia gravis in an initially asymptomatic mother. Neuromusc Disord 5:59-65, 1995.

10. Bayne LG: Hand assessment and management in arthrogryposis multiplex congenita. Clin Orthop 194: 68-73, 1985.

11. Beals RK: The distal arthrogryposes: a new classification of peripheral contractures. Clin Orthop 435: 203-210, 2005.

12. Bernstein RM: Arthrogryposis and amyoplasia. J Amer Acad Orthop Surg. 10: 417-424, 2002.

13. Bevan WP, Hall JG, Bamshad M et al: Arthrogryposis multiplex congenita (amyoplasia): an orthopaedic perspective. J Pediatr Orthop 27: 594-600, 2007.

14. Brandt S: A case of arthrogryposis multiplex congenita. Anatomically appearing as foetal spinal muscular atrophy Act Paediatr 34: 365-370, 1947.

15. Carlson WO, Speck GJ, Vicari V, Wenger D: Arthrogryposis multiplex congenita: a long term follow-up study. Clin Orthop 194: 115-123, 1985.

16. Darin N, Kimber E, Kroksmark AK, Tulinius M: Multiple congenital contractures: birth prevalence, etiology and outcome. J Pediatr 140: 61-67, 2002.

17. Denecke J, Zimmer KP, Kleta R et al: Arthrogryposis, renal tubular dysfunction, cholestasis (ARC syndrome): case report and review of the literature. Klin Padiatr 212: 77-80, 2000.

18. Eastham KM, McKiernan PJ, Milford DV et al: ARC syndrome: an expanding range of phenotypes. Arch Dis Child 85: 415-420, 2001.

19. Friedlander JL, Westin GW, Wood WL: Arthrogryposis multiplex congenita J Bone Joint Surg 50A: 89-112, 1968.

20. Gibson DA, Urs NDK: Arthrogryposis multiplex congenita J Bone Joint Surg 52B:483-493, 1970.

21. Green ADL, Fixsen J, Lloyd-Roberts GC: Talectomy for arthrogryposis multiplex congenita. J Bone Joint Surg 66B: 697-699, 1984.

22. Guerin JR: Recherches sur lis deformities congenitales cuez les monsters. Paris, 1880.

23. Guidera KJ, Drennan JC: Foot and ankle deformities in arthrogryposis multiplex congenita Clin Orthop 194: 93-98, 1985.

24. Guimaraes AS, Marie SK: Dominant form of arthrogryposis multiplex congenita with limited mouth opening: a clinical and imaging study. J Orofac Pain 19: 82-88, 2005.

25. Hahn G: Arthrogryposis. Pediatric review and rehabilitative aspects. Clin Orthop 104: 104-114, 1985.

26. Hall JG: Analysis of Pena Shokeir phenotype. Am J Med Genet 25:99-117, 1986.

27. Hall JG: Arthrogryposis multiplex congenita: etiology, genetics, classification, diagnostic approach and general aspects. J Pediatr Orthop B 6: 159-166, 1997.

28. Hall JG: Genetic aspects of arthrogryposis. Clin Orthop 194: 44-53, 1985.

29. Heffez L, Doku HC, O'Donnell JP: Arthrogryposis multiplex complex involving the temporomandibular joint. J Oral Maxillofac Surg 43: 539-542, 1985.

30. Hoff JM, Dalveit AK, Gilhus NE: Arthrogryposis multiplex congenita—a rare fetal condition caused by maternal myasthenia gravis. Acta Neurol Scand Suppl 183: 26-27, 2006.

31. Jaber L, Weitz R, Bu X: Arthrogryposis multiplex congenita in an Arab kindred: an update. Am J Med Genet 55:331-334, 1995.

32. Kang PB, Hart HGW, David WS et al: Diagnostic value of electromyography and muscle biopsy in arthrogryposis multiplex congenita. Ann Neurol 54: 790-795, 2003.

33. Kobayashi H, Baumbach L, Matise TC et al: A gene for a severe lethal form of X-linked arthrogryposis (X-linked infantile spinal muscular atrophy) maps to human chromosome Xp11.3-q11.2. Hum Mol Genet 4: 1213-1216, 1995.

34. Krakowiak PA, O'Quinn JR, Bohnsack JF et al: A variant of Freeman-Sheldon syndrome maps to 11p15.5pter. Am J Hum Genet. 60: 426-432, 1997.

35. Kroksmark AK, Kimber E, Jerre R et al : Muscle involvement and motor function in amyoplasia. Am J Med Genet. 140:1757-1767, 2006.

36. Laureano AN, Ryback LP: Severe otolaryngologic manifestions of arthrogryposis multiplex congenita. Ann Otol Rhinol Laryngol 99: 94-97, 1990.

37. Lewin P: Arthrogryposis multiplex congenita J Bone Joint Surg 7A: 630-638, 1925.

38. Livingstone IR, Sack GH Jr.: Arthrogryposis multiplex congenita occurring with maternal multiple sclerosis. Arch Neurol 41: 1216-1217, 1984.

39. Lloyd-Roberts GC, Lettin AWF: Arthrogryposis multiplex congenita. J Bone Joint Surg 52B: 494-508, 1970.

40. Mayhew IG: Neuromuscular arthrogryposis multiplex congenita in a thoroughbred foal. Vet Pathol 21: 2: 187-192, 1984.

41. Mennen U, van Heest A, Ezaki MB et al: Arthrogryposis multiplex congenita. J Hand Surg 30: 468-474, 2005.

42. Middleton DS: Studies on prenatal lesions of striated muscle as a cause of congenital deformity. Edinburgh Med J 401, 1934.

43. Murray C, Fixsen JA: Management of knee deformity in classical arthrogryposis multiplex congenita (amyoplasia congenita). J Pediatr Orthop B 6: 186-191, 1997.

44. Narkis G, Landau D, Manor E et al: Genetics of arthrogryposis: linkage analysis approach. Clin Orthop 456: 30-35, 2007.

45. Niki J. Staheli LT, Mosca VS: Management of club foot deformity in amyoplasia. J Pediatr Orthop 17: 803-807, 1997.

46. Obarski TP, Fardal PM, Bush CR, Leier CV: Stenotic aortic and mitral valves in three adult brothers with arthrogryposis multiplex congenita. Am J Cardiol 96: 464-466, 2005.

47. Otto AW: Monstrum humanum extremitatibus incurvatus. Monstrorum sexcentrorum descriptio anatomica inVratislaviae Museum. Anatomico-Pathologicum, Breslau 321-322, 1841.

48. Pakkasjarvi, N, Ritvanen A, Herva R et al: Lethal congenital contracture syndrome (LCCS) and other lethal arthrogryposes in Finland—an epidemiological study. Am J Med Genet A 140A: 1834-1839, 2006.

49, Pena, SD, Shokeir MH: Syndrome of camptodactyly, multiple ankyloses, facial anomalies and pulmonary hypoplasia: a lethal condition. J Pediatr 85:373-375, 1974.

50. Polizzi A, Huson SM, Vincent A: Teratogen update: maternal myasthenia gravis as a cause of congenital arthrogryposis. Teratology 62: 332-341, 2000.

51. Quinn CM, Wigglesworth JS, Heckmatt J: Lethal arthrogryposis multiplex congenita: a pathologic study of 21 cases. Histopathology 19:155-162, 1991.

52. Rocher HL: Les raideurs articularies congenitales multiples. J Med Bordeaux 43: 722-780, 1913.

53. Sakai T, Kikuchi F, Takashima S et al: Neuropathological findings in the cerebro-oculo-facial-skeletal (Pena-Shokeir II) syndrome. Brain Dev 19: 58-62, 1997.

54. Sarwak JF. Macewen GD, Scott CI: Amyoplasia (a common form of arthrogryposis). J Bone Joint Surg 72A: 465-469, 1990.

55. Schanz A: Ein fall von multiplen congenitalen contracturen. Z Orthop 5: 9-, 1897.

56. Sells JM, Jaffe KM, Hall JG: Amyoplasia, the most common type of arthrogryposis: the potential for good outcome. Pediatrics 97: 225-231, 1996.

57. Sheldon W: Amoplasia congenital (multiple congenita articular rigidity: arthrogryposis multiplex congenita). Arch Dis Child 7: 117-136, 1932.

58. Shohat, M, Lotan R, Magal N et al: A gene for arthrogryposis multiplex congenita neuropathic type is linked to D5S394 on chromosome 5qter. Am J Hum Genet 61: 1139-1143, 1997.

59. Staheli LT, Chew DE, Elliot JS, Mosca VS: Management of hip dislocations in children with arthrogryposis. J Pediatr Orthop 7: 681-685, 1987.

60. St. Clair HS, Zimbler S: A plan of management and treatment results in the arthrogrypotic hip. Clin Orthop 194: 74-80, 1985.

61. Steinberg B, Nelson VS, Feinberg S et al: Incidence of maxillofacial involvement in arthrogryposis multiplex congenita. J Oral Maxillofac Surg 54: 956-959, 1996.

62. Stern WG: Arthrogryposis multiplex congenita. JAMA 81: 1507-1510, 1923.

63. Strehl E, Vanasse M, Brochu P: EMG and needle muscle biopsy studies in arthrogryposis multiplex congenita. Neuropediatrics. 16: 225-227, 1985.

64. Swinyard CA, Bleck EE: The etiology of arthrogryposis (multiple congenital contracture). Clin Orthop 194:15-29, 1985.

65. Tanamy MG, Magal N, Halpern GJ et al: Fine mapping places for the gene for arthrogryposis multiplex congenita neuropathic type between D5S394 and D5S2069 on chromosome 5qter. Am J Med Genet. 104: 152-156, 2001.

66. Thompson GH, Bilenker RM: Comprehensive management of arthrogryposis multiplex congenita. Clin Orthop 194: 6-14, 1985.

67. Williams PF: Management of upper limb problems in arthrogryposis. Clin Orthop 194: 60-67, 1985

68. Wolf A, Roverud E, Poser C: Amoplasia congenita Neuropathol Exp Neurol 14:112-, 1955.

69. Yau PWP, Cow W, Li YH et al: Twenty-five year followup of hip problems in arthrogryposis multiplex congenita. J Pediatr Orthop 22: 359-363, 2002.

CHAPTER 6

Chondrodysplasia Punctata . . . Several Forms of a Rare Disorder

Introduction:

Chondrodysplasia punctata, as the name implies, describes a number of disorders in which the epiphyses become "stippled" with calcium. The patients are of short stature and the long bones are diminished in length and have punctate calcific deposits in the cartilaginous portions of the skeleton. Cranial facial defects and cataracts are often present. There are five separate entities which have the name chondrodysplasia punctata and depending on the type of disease, it may only occur in males; may be associated with severe early mental impairment; may have foot deformities and contractures; have severe skin abnormalities; and may be characterized by early death. It seems unreasonable in a chapter of this sort to review and describe each of the five disorders reported in the literature since most of them are rare and have very limited orthopaedic presentations. Instead it seems appropriate to confine the discussion to the most common and most characteristically orthopaedic disorder, known as Conradi-Hünermann disease or Conradi-Hünermann-Happle disease and is otherwise known as CDPX2. The disorder is X-linked dominant and is considerably more common in females.

Nomenclature and history:

The type of chondrodysplasia punctata which will be discussed in this chapter is also known as Conradi-Hünermann disease; Conradi-Hünermann-Happle disease chondroangiopathia calcarea seu punctata; chondrodysplasia calcificans congenita; epiphysealis punctata dominant type; chondrodystrophia calcificans; stippled epiphyses; and punctate epiphyseal dysplasia.

Erich Conradi first described the disease in 1914 (8) and Carl Hünermann further contributed in a report in 1931 (26). In subsequent years a number of authors contributed to the knowledge of the bone and epiphyseal cartilage disorders, the visual disturbances, dermatologic problems and the facial anomalies. These included Bloxsom and Johnston who in a publication in 1938 (5) disclosed some of the unusual features of the disorder including the early death for some patients. In 1954, Paul (39) described some of the unique X-ray changes. Comings and coworkers in 1968 (7) and Gwinn and Lee in 1971 (17) both further clarified the unusual and sometimes variable presentation of the disease. Bodian in 1966 and Elidin et al in 1977 (12) described the dermatologic features. Sheffield and coworkers in 1976 described the clinical features of the disorder. Spranger et al in 1979 (47) and Manzke and coworkers in 1980 (34) described some of the remarkable clinical aspects and reviewed the heterogeneity of the disorder.

It was Rudolf Happle, a competent geneticist and his coworkers who beginning in 1977, identified the disorder as X-linked and expanded knowledge as to its clinical characteristics (21). They described the female predominance of the disorder (19,20,21) and also indicated the difference between this entity and some of the others, especially the rhizomelic disorder with the early death rate and severe mental impairment (19). Because of Dr. Happle's commitment to the subject, the disease became known as Conradi-Hünermann-Happle disease.

Biologic and genetic characteristics of Conradi-Hünermann-Happle disease:

Histologic studies by Pazzaglia et al (40) and Hoang and coworkers (24) in recent years described the nature of the cartilage calcific stippling and suggested that it was related to unusual sterol substances within the tissues. Laboratory studies on patients with CDPX2 show increased amounts of 8-dehydrocholesterol and 8(9)-cholesterol in the plasma and as a result, the serum cholesterol levels are usually low (11,13,31,32,36). Molecular studies have shown that the disorder is caused by mutations in the human emopamil-binding protein (EBP) which is mapped to Xp11.22-p11.23 (2,22,27,33,48). The gene comprises five exons, four of which encode a 230-amino-acid, four transmembrane domain protein that functions as a 3beta-hydroxysteroid Delta(8)-Delta(7)-isomerase in the final steps of cholesterol biosythesis (3,6). Plasma sterol alterations of this sort cause a sometimes profound lowering of the serum cholesterol concentration and have been suspected as being the cause of the clinical characteristics of the syndrome; but clinical administration of cholesterol does not seem to improve the patient's problems (11,13,36). Males with the disorder are less commonly encountered and are more likely to have mental problems and an early demise (9,10). Females with the disorder seem to do better, possibly because of sometimes a marked increase in additional mutations which improve their mentation and increase their survival (23,50).

Clinical presentation of Conradi-Hünermann-Happle disease:

As noted above, the clinical syndrome of Conradi-Hünermann-Happle disease is considerably more common in females who have a less severe disorder than males (28, 43). Female children with the disease usually have normal mentation. They are often short statured and have asymmetrical limb shortening (20,23,28,37,43,45,46). The bones that are excessively short include the femora, tibiae and humeri (1,16,28,38). Flexion contractures of the hips and knees are commonly present and at times disabling (43,45). There are also some remarkable deformities of the pelvis with defective acetabulae and pubic bones (1,43,45). Scoliosis is common but is usually mild and non-progressive but may be a problem, particularly if it is present in the cervical region (15,39,49). Abnormal foot bones result in talipes equinovarus (35,38,43). Hip dislocations occur later in life and genu valgum is commonly present (38,43). Mild nail changes including platonychia or onycoschisis are frequently present (28,35).

Affected children have sparse, coarse hair. The face has frontal bossing which causes a distinctive appearance known as "koala bear facies" (4,12,14,25). The nasal bone is hypoplastic and the face is often asymetrical in appearance (25,45). The patients have difficulty with vision, often related to early cataracts (19,42,43). Over time, many of the patients develop micropthalmus, nystagmus, glaucoma and optic nerve atrophy. All of these most often lead to blindness in the early teens (42,43). The teeth are usually normal.

Cutaneous problems are frequently present early in the illness. These may consist of a cutaneous ichthyosis with adherent large scales (4,14,18). The skin changes tend to lessen in severity as the patient ages but patchy alopecia with coarse sparse hair remains as a disturbing finding.

Male children with the disorder are less frequently seen but have much more severe problems including markedly impaired mentation, severe skin problems, early cataract-caused blindness. The disorder almost always causes death at an early age, often related to pulmonary complications (9,10).

Radiographic imaging show punctuate calcification of the cartilaginous portions of the skeleton including both ends of the long bones, tarsal bones, vertebra adjacent to the disk spaces and ischial bones (1,28,38,43),. The calcification disappears later in life. The asymmetric shortening of the long bones is characteristic and calcification may be present in the soft tissues surrounding joints and epiphyseal centers.

As the X-linked female patients grow their mental abilities generally remain normal but they are short statured, and often have difficulty with vision and sometimes remarkably impaired bones and joints.

Treatment of patients with Conradi-Hünermann-Happle disease:

In view of the low levels of blood cholesterol in patients with CDPX2, several investigators have suggested that cholesterol administration would perhaps be helpful (11,13,31,36). Regrettably, there is no evidence to support this approach. Furthermore, no genetic or biologic treatment protocol has emerged. As far as the patient's skin is concerned, the usual dermatological agents can be applied and are sometimes helpful (4,12,18,29). The skin problem however improves with advancing age.

Opathalmologic treatment is essential. Cataracts may be removed and other aspects of the eye problems can be dealt with in standard fashion, but often with limited success (19,28,42).

Bone problems are sometimes major issues (1,15,28). Treatment of fractures is often necessary and dislocations of joints may require reconstruction as the patient ages. Osteotomies of the deformed long bones are sometimes essential and rapidly advancing scoliosis, especially of the cervical region may require spinal corrective efforts (15,30,44,49).

It seems essential for the parents of an affected child to have genetic counseling. Having more than one child with this disorder can be very destructive to families. Currently it is possible to make a prenatal diagnosis of an affected child (41). It seems hopeful that some reasonable biologic solution will be produced and utilized for the patients and their affected parents and siblings in order to decrease the likelihood of further occurrences of this bizarre disorder or some of the other similar and sometimes even more severe diseases.

Discussion:

Conradi-Hünermann-Happle disease is rarely encountered by orthopaedic surgeons and most have no idea what the problems are for the patients and their families. The disease is X-linked dominant and in families that have the gene error, female children appear with skin problems, facial abnormalities, short stature, scoliosis, bone shortening, deformities, epiphyseal abnormalities and most often sometimes severe ocular problems which cause blindness at an early age. Even sadder, males who develop the disorder usually die within a few years after birth. Although we now know a great deal about the genetic causation of the disease and specifically its relationship to cholesterol production and action on the tissues, we currently have no way of adding normal cholesterol as a treatment and even more disturbing, we do not have a method of altering the genetic error with appropriate therapy.

It seems logical to continue studies of the genetic problems and even treat the children while they are in utero. It seems reasonable to continue to perform research in this crucial area. If some success occurs, there is no doubt that the affected patients, their parents and their treating physicians will be eternally grateful.

References:

1. Andersen PE Jr., Justensen P: Chondrodysplasia punctata. Report of two cases. Skeletal Radiol 16: 223-22, 1987.
2. Ausevarat S, Tanpaibood P, Tongkobpetch S et al: Two novel EBP mutations in Conradi-Hünermann-Happle syndrome. Eu J Dermatol 18: 391-393, 2008.
3. Becker K, Csikos M, Horrvath A, Karpati S: Identification of a novel mutation in 3beta-hydroxysteroid-Delta8-Delta7-isomerase in a case of Conradi-Hünermann-Happle syndrome. Exp Dermatol 10: 286-289, 2001.
4. Bodian EL: Skin manifestations of Conradi's disease: chondrodystrophia congenita punctata. Arch Dermatol 94: 743-748, 1966.
5. Bloxsom A, Johnston RA: Calcinosis universalis with unusual features. Am J Dis Child 56: 103-109, 1938.
6. Braverman N, Lin P, Moebius FF et al: Mutations in the gene encoding 3 beta hydroxysteroid-delta 8, delta 7-isomerase cause x-linked Conradi-Hünermann syndrome Nat Genet 22: 291-294, 1999.
7. Comings DE, Papazian c, Schoene HR: Conradi's disease: chondrodystrophica calcificans congenita, congenital stipple epiphysis. J Pediatr 72:h 63-69, 1968.

8. Conradi E: Vorzeitiges auftreten von knochen und eigenartigen verkalkungskernan bei chondrodystrophia foetalis hypoplastica. Jahrb Kinderh 80: 86-97, 1914.

9. Crovato F, Rebora A: Acute manifestations of Conradi-Hünermann syndrome in a male adult. Arch Dermatol 121: 1064, 1065, 1985.

10. De Raeve L, Song M, De Dobbeleer G et al: Lethal course of X-linked dominant chondrodysplasia punctata in a male newborn. Dermatologica 178: 167-170, 1989.

11. DiPreta EA, Smith KJ, Skelton H: Cholesterol metabolism defect associated with Conradi-Hünermann-Happle syndrome. Int J Dermatol39: 846-850, 2000.

12. Elidin DV, Esterly NB, Barnzai AK et al: Chondrodysplasia punctata (Conradi-Hünermann syndrome) Arch Dermatol 113: 1431-1434, 1977.

13. Farese RV Jr., Heerz J: Cholesterol metabolism and embyrogenesis. Trends Genet 14: 115-120, 1998.

14. Feldmeyer L, Mevorah B, Grzeschik KH et al: Clinical variation in X-linked dominant chondrodysplasia punctata (X-linked dominant ichthyosis). Br J Dermatol 154: 766-769, 2006.

15. Garnier A, Dauger S, Eurin D et al: Brachytelephalangic chondrodysplasia punctata with severe spinal cord compression: report of four new cases. Europ J Pediatr 166: 327-331, 2007.

16. Gobello T, Mazzanti C, Fileccia P et al: X-linked dominant chondrodysplasia punctata (Happle syndrome) with uncommon symmetrical shortening of the tubular bones. Dermatology 191: 323-327, 1995.

17. Gwinn JL, Lee FA: Conradi's disease (chondrodystrophia calcificans congenita) J Pediatr 63: 63-69, 1971.

18. Hamaguchi T, Bondar G, Siegfried E et al: Cutaneous histopathology of Conradi-Hünermann syndrome. J Cutan Pathol 22:38-41, 1995.

19. Happle R: Cataracts as a marker of genetic heterogeneity in chondrodysplasia punctata. Clin Genet. 19: 64-66, 1981.

20. Happle R: X-linked dominant chondrodysplasia punctata: review of the literature and report of a case. Hum Genet 53: 65-73, 1979.

21. Happle R, Matthass H-H, Macher E: Sex-linked chondrodysplasia punctata. Clin Genet 11: 73-76, 1977.

22. Has C, Bruckner-Tuderman L, Müller D: The Conradi-Hünermann-Happle syndrome (CDPX2) and emopomil binding protein: novel mutations and somatic and gonadal mosaicism. Hum Mol Gene 9: 1951-1955, 2000.

23. Herman GE, Kelley RI, Pureza V et al: Characterization of mutations in 22 females with X-linked dominant chondrodysplasia punctata (Happle syndrome). Genet Med 4: 434-438, 2002.

24. Hoang MP, Carder KR, Pandya AG, Bennett MJ: Icthyosis and keratotic follicular plugs containing dystrophic calcification in newborns: distinctive histopathologic features of X-liked dominant chondrodysplasia punctata (Conradi-Hünermann-Happle syndrome). J Dermatopathol 26; 53-58, 2004.

25. Hochman M, Fee WEJ: Conradi-Hünermann-syndrome case report. Ann Otol, Rhinol Laryngol 96: 565-568, 1987.

26. Hünermann C: Chondrodystrophia calcificans congenita als abortive form der chondrodystrophie. Zschr Kinderh 51: 1-19, 1931.

27. Ikegawa S, Ohashi H, Ogata T et al: Novel and recurrent EBP mutations in X-linked dominant chondrodysplasia punctata Am J Med Genet. 94: 300-305, 2000.

28. Irving MD, Chitty LS, Mansour S, Hall CM: Chondrodysplasia punctata: a clinical diagnostic and radiologic review. Clin Dysmorphol 17: 229-241, 2008.

29. Kalter DC, Atherton DJ, Clayton PT: X-linked dominant Conradi-Hünermann syndrome presenting as congenital erythroderma. J Am Acad Derm 21: 248-256, 1989.

30. Kaufmann JH, Mahboubi S, Spackman TJ et al: Tracheal stenosis as a complication of chondrodysplasia punctata. Ann Radiol 19: 203-209, 1976.

31. Kelley RI, Wilcox WG, Smith M et al: Abnormal sterol metabolism in paitents with Conradi-Hünermann-Happle syndrome and sporadic lethal chondrodysplasia punctata. Am J Med Genet 83: 213-219, 1999.

32. Kolb-Maurer A, Grzeschik K-H, Haas, D et al: Conradi-Hünermann-Happle syndrome (X-linked dominant chondrodysplasia punctata) confirmed by plasma sterol and mutation analysis. Acta Derm Venereol 88: 47-51, 2008.

33. Lo IFM, Kwong W, Lee RSY et al: A novel mutation of the EPB gene causes Conradi-Hünermann syndrome. HK J Paediatr 11: 327-330, 2006

34. Manzke H, Cristophers E, Wiedemann HR: Dominant sex-linked inherited chondrodysplasia punctata: a distinct type of chondrodysplaia punctata. Clin Genet 17: 97-107, 1980.

35. Maroteaux P: Brachytelephalangic chondrodysplasia punctata: a possible X-linked recessive form. Hum Genet 82: 167-170, 1989.

36. Martanova H, Krepelova A, Baxova A et al: X-linked dominant chondrodysplasia punctata (CDPX2): multisystemic impact of the defect in cholesterol biosynthesis. Prague Med Rep 108: 263-269, 2007.

37. Meuller RF, Crowle PM, Jones RA et al: X-linked dominant chondrodysplasia punctata. Am J Med Genet 20137-144, 1985.

38. Netter FH: Dysplasia of bone and soft tissues. Musculoskeletal System Part II. Volume 8. Summit, NJ, Ciba-Geigy Corporation, 1990, page 10.

39. Paul LW: Punctate epiphyseal dysplasia. Am J Roentgen. 71: 941-946,1954.

40. Pazzaglia UE, Zarattini G, Donzelli C et al:The nature of cartilage stippling in chondrodysplasia punctata: hstopathological study of Conradi-Hůnermann syndrome. Fetal Pediatr Pathol 27:71-81, 2008.

41. Pryde PG, Bawle E, Brandt F et al: Prenatal diagnosis of nonrhizomelic chondrodysplasia punctata (Conradi-Hünermann-Happle syndrome). Am J Med Genet. 47: 426-431, 1993.

42. Ramkisson YD, Mayer EJ, Gibbon C, Haynes RJ: Vitreoretinal abnormalities in the Conradi-Hünermann form of chondrodysplasia punctata. Br J Opthalmol 88: 973-974, 2004.

43. Rimoin DL, Lachman RS: Genetic disorders of the osseous skeleton. in McKusick's Heritable Disorders of Connective Tissue. Fifth Edition, Beighton, P editor. St Louis Mosby, 1993, 625-627.

44. Sheffield LJ, Danks DM, Mayne V, Hutchinson LA: Chondrodysplasia punctata: 23 cases of a mild and relatively common variety. J Pediatr 89: 916-923, 1976.

45. Sheffield LJ, Halliday JL, Danks DM et al: Clinical, radiological and biochemical classification of chondrodysplasia punctata. Am j Med Gent 45 (Suppl) A64, 1989.

46. Silengo MC, Luzzatti L, Silverman FN: Clinical and genetic aspects of Conradi-Hünermann disease: a report of three familial cases and review of the literature. J Pediatr 97: 911-917, 1980.

47. Spranger JW, Opitz JM, Bidder Ve: Heterogeneity of chondrodysplasia punctata. Humangenetik 11: 190-212, 1979.

48. Steijlen PM, van Geel M, Vreeburg D et al: Novel EBP gene mutations in Conradi-Hünermann-Happle syndrome. Brit J Dermatol 157: 1225-1229, 2007.

49. Violas P, Fraisse B, Chapuis M, Bracq H: Cervical spine stenosis in chondrodysplasia punctata. J Pediatr Orthop 16: 443-445, 2007.

50. Whittock NV, Izatt L, Mann A et al: Novel mutations in X-linked dominant chondrodysplasia punctata (CDPX2). J Invest Dermatol 121: 939-942, 2003.

CHAPTER 7

Cleidocranial Dysotosis: A Rare Autosomal Dominant Disorder

Cleidocranial dysostosis is an unusual genetic disorder with some remarkable characteristics. Despite its autosomal dominant nature, the disease is uncommon and does not show any gender or ethnic prevalences. Delayed ossification of the calvarium leads to excessively large fontanels, facial and oral abnormalities. The clavicles are either absent, fractured or reduced in size and as a result the anterior range of shoulder movements is extraordinary. The patients often have delayed tooth eruption with short thin roots and sometimes a remarkable number of supernumerary teeth. Additional less common osseous defects include deformed and dislocated hips, narrow pelvis, funnel chest, deformed hands and joint hypermobility. The patients are sometimes of slightly shortened stature and have normal mentation.

Nomenclature and history:

Other terms used to describe the cleidocranial dysostosis syndrome include Marie-Sainton syndrome, Scheuthauer-Marie syndrome, Scheuthauer-Marie-Sainton syndrome, cleidocranial dysplasia, cleidocranial digital dysostosis, dysostosis cleidocraniopelvina, dysostosis generalisata, mutational dysostosis, osteodental dysplasia and pelvicocleidocranial dysostosis.

The abnormal disorder of the cranium and clavicles was first described by Martin in 1765 (26) and by Morand in 1776 (32), but was much more clearly defined by Scheuthauer in 1871 (38) and Barlow in 1883 (2). It was Pierre Marie and Paul Sainton (25) however who in 1898 not only reported several cases but also named the disease "cleidocranial dysostosis". Reports by Carpenter (5) and Terry (47) both in 1899 emphasized the structural abnormality or complete absence of clavicles in affected patients. In 1916, Langmead (23) reported 18 cases in four generations. In 1923, McCurdy and Baer (28) described 9 patients in three generations, and together these studies fully established the autosomal dominant nature of the disorder. A report by Earl McBride in 1927 (27) further emphasized the clavicular abnormalities associated with the disease and Kelley in an article in 1929 (20) described the frequently encountered hereditary absence of the clavicle in patients with the disorder. In 1929 in an extraordinary publication by Seth M. Fitchet (10) in the Journal of Bone and Joint Surgery, he not only described the entity, but provided 113 literature references regarding the disorder. This article provides great detail as to the nature and extent of the orthopaedic difficulties with emphasis on the calvarium, the facial structures, the hands and especially the clavicles. In 1951, Cole and Levin (6) described the radiologic changes in patients with cleidocranial dysostosis with emphasis on the skull and clavicles. Another unusual description of the disease was by Jackson in 1951 (15) who traced a large numbers of members of a family named Arnold. His article describes the fact that 70 of these had calvarial deformity which he termed the "Arnold Head". He attributed much of it to changes in the bony structure findings as previously reported by Eltorm in 1945 (9), Thomsen and Guttadauro in 1952 (48), Wee and Pillay in 1964 (52) and Jarvis and Keats in 1974 (16). Another remarkable effort was that of Philip Koch and Wade Hammer (22), who in 1978, provided an extraordinary assessment of oral and dental problems; and along with other investigators, noted that one of the striking changes in the dental structure was the loss of cementum (37,42,52).

Biologic aspects of cleidocranial dysostosis:

Based on remarkable genetic investigation in 1997 and 1999 by Stefan Mundlos (33,34) and other subsequent further studies by others (7,12,24,31,35,44,46,54) the gene for cleidocranial dysostosis was clearly identified. It consists of a core binding factor alpha-1 (CBFA1) otherwise more frequently known as RUNX2 which is located on the short arm of chromosome 6p214, t(6;18) (p12;q24) (7,12). RUNX2 encodes a transcription factor that activates osteoblast differentiation and is composed of nine exons. Insertions, deletions, nonsense and missense mutations have been identified in families with cleidocranial dysostosis (24,31). The genetic error causes wormian bones and a delay in ossification during childhood which may result in multiple structural errors in the calvarium, mandible and clavicles with age (44). In addition RUNX2 is a major regulator of chondrocyte differentiation and can cause epiphyseal abnormalities which are responsible for some of the alterations in the structure of the hands, the mandible, the spine and the lower extremities in the affected patients (53). Growth is usually not affected so that most of the patients are of normal stature (7,16,31,40). Mentation is not impaired (7,10,40). Diminished quantities of cementum are in large measure responsible for the anomalies of the teeth, which at times can be quite striking (39,52).

Clinical presentation of cleidocranial dysostosis:

Many patients with cleidocranial dysostosis have only limited findings and are not known to have the disease. The patients are rarely shortened in height and their mental status is reported to be normal (7,10,30,40). One of the methods of discovery is related to the autosomal dominant status so that the only feature which is likely to bring the disease to light is the family history. In a series by Cooper et al, (7) involving 116 patients with the disorder, there were only 53 patients with known disease and 56 related patients who were not known to have the disease but were genetically positive.

Clinical skeletal problems are the most common. Genu valgum and pes planus are present in almost all of the patients (1,7,16). Multiple fractures occur quite regularly in affected patients and can be disabling (43). Healing of fractures is relatively slow and those affecting the clavicles may not heal at all (48). Clavicular absence or reduction in size is a frequent finding and hypermobile shoulders can be striking (1,7,14,16,27,40). Some of the patients can bring their shoulders together to touch anteriorly. Scoliosis is present in approximately 20% of the patients but is usually not severe and may be asymptomatic (7,10,16). Other skeletal problems include joint laxity and pain especially in the knees and hips, elbow dislocations, short digits in hands and feet and hyperplastic nails (7,43). The pelvis is affected and on X-ray is shown to have delayed ossification of the pubis with a wide pubic symphysis, hypoplasia of the iliac wings, widening of the sacroiliac joints and an enlarged femoral neck (1,6,7,10,40,). An occasional finding on imaging of the hand is accessory epiphyses on the metacarpal bones (7,31). The face and head are often involved but the findings may be somewhat subtle (1,3,7). The head is large and out of proportion to the bones of the face (13,17,19,31). Bulging forehead and prominent cranial bossing are usually present. Broad sutures and large persistent fontanelles can change the shape of the calvarium in infants (18). The eyes are widely spaced and the nasal bridge may be flat (1,13,17.40,41,43). Hearing loss is frequent (51). Occasional additional neurologic problems may occur (21,30,50).

Dental problems may be the most striking difficulties encountered by the affected patients (3,10,22,31,39). The teeth fail to erupt based in large measure on the absence of cellular cementum and an increase in the amount of acellular cementum surrounding the roots of the teeth (42,52). One often encounters supernumerary or unerupted teeth and the deciduous teeth are retained for a prolonged period (3,8.40,45). The maxilla may be underdeveloped and narrow resulting in a mandibular prominence. The mandibular symphysis may remain open even in adulthood (7,8).

It is important to note that some other disorders may bear some resemblance to cleidocranial dysostosis, especially in terms of calvarial and facial abnormalities (36). These include Crane-Heise syndrome, mandibuloacral dysplasia, pycnodysostosis, Crouzon's disease, hypophosphatasia and hypothyroidism. If there is any doubt analysis for the RUNX2 gene is appropriate (7).

Treatment of patients with cleidocranial dysostosis:

It is essential to note that many patients with this disorder have such mild symptomatology that they are not recognized as having problems. There is only infrequent shortness of stature and no mental abnormality or early demise. Patients may be involved in a series of normal life activities including education, employment, marriage, family development and participation in social and economic programs. For these patients, since there is no chemical or genetic therapy available, even when the disease is identified, there is no reason to treat them unless they develop some symptomology.

Some of the problems which require treatment include those affecting the calvarium, face and oral cavity (4,29,49). Deafness is common and the use of hearing aids is sometimes essential. Visual loss is difficult to treat but fortunately is most often mild. Nasal narrowing which can result in frequent sinus infections can sometimes be treated with minor surgery (7). Oral problems often require extensive treatment to remove erupted, impacted or excessive numbers of deciduous teeth (4,11,45,51). Imaging is very helpful and should be repeated regularly to be certain that the problems are under control (16,48).

Treatment of bone problems is sometimes necessary (10,40,43). For patients with structural malalignment, osteotomy and insertion of appropriate hardware is sometimes important to maintain proper limb function (7,16,43). Dislocating joints must be realigned and possibly will require insertion of artificial joint hardware. Treatment of the absence or marked shortening of the clavicle is usually unnecessary as is the management of asymptomatic pelvic malalignment or mild scoliosis.

Conclusions:

In many ways, cleidocranial dysostosis is an unusual genetic disorder. It is autosomal dominant and thus has a high rate of occurrence in families, but because for many patients, the symptoms are mild, it is not always recognized in siblings. In terms of the nature of the disorder, the most remarkable features include oral problems and absence or shortening of the clavicles. The clavicular abnormalities often result in a remarkable structural alteration of the shoulders and upper body. Patients may require extensive oral surgery and even then may have difficulties. The orthopaedic issues are principally associated with extremity malalignment, hand and foot structural abnormality and increased risk of fractures.

No chemical or biologic treatment is currently available although it is possible that with development of gene alteration technology, the use of some sort of agent to interfere with the actions of RUNX2 (CBFA1) could help to eliminate the problem.

References:

1. Alves N, de Oliveira R: Cleidocranial dysplasia-a case report. Int J Morphol. 26: 1065-1068, 2008.
2. Barlow T: Congenital absence of both clavicles and malformation of cranium. Br Med J. 1: 909, 1883.
3. Becker A, Lustmann J, Shteyer A: Cleidocranial dysplasia; part 1. general principles of the orthodontic and surgical modality. Am J Orthod Dentofac Orthop 111: 28-33, 1997.
4. Becker A, Shteyer A, Bimstien E, Lustmann J: Cleidocranial dysplasia; part 2. treatment protocol for the orthodontic and surgical modality. Am J Orthod Dentofac Orthop 111: 173-183, 1997.
5. Carpenter G: A case of absence of the clavicles. Lance 1: 13-16, 1899.
6. Cole WR, Levin S: Cleidocranial dysostosis. Br J Radiol 24: 549-555, 1951.
7. Cooper SC, Flaitz CM, Johnston DA et al: A natural history of cleidocranial dysplasia. Am J Med Genet 104: 1-6, 2001.
8. Douglas BL, Greene HJ: Cleidocranial dysostosis: report of a case. J Oral Surg 28: 41-43, 1969.
9. Eltorm H: Case of cleidocranial dysostosis. Acta Radiol 26: 69-75, 1945.
10. Fitchet SM: Cleidocranial dysostosis: hereditary and familial. J Bone Joint Dis 11: 836-866, 1929.
11. Frohberg U, Tiner BD: Surgical correction of facial deformities in a patient with cleidocranial dysplasia J Craniofac Surg 6: 49-53, 1995.

12. Gelb BD, Cooper E, Shivell M, Desnick RJ: Genetic mapping of cleidocranial dysplasia locus on chromosome band 6 and 6p21 to include microdeletion. Am J Med Genet 58: 200-205, 1995.

13. Golan I, Waldeck A, Baumert U et al: Anomalies of the skull in cleidocranial dysplasia. HNO 52: 1061-1066, 2004.

14. Issever AS, Diederichs G, Tuischer J: Diagnosis of cleidocranial dysplasia in routine chest radiograph. Circulation 116: e116-e118, 2007.

15. Jackson WPU: Osteo-dental dysplasia (cleido-cranial dysostosis); the "Arnold head". Acta Med Scand 139:292-307, 1951.

16. Jarvis JL, Keats TB: Cleidocranial dysostosis: a review of 40 cases. AJR121: 5-16, 1974.

17. Jensen BL: Cleidocranial dysplasia: craniofacial morphology in adult patients. J Craniofac Genet Devel Biol 14: 163-176, 1994.

18. Jensen BL, Kreiborg S: Development of the skull in infants with cleidocranial dysplasia. J Craniofac Genet Devel Biol 13: 69-97, 1993.

19. Jensen BL, Kreiborg S: Craniofacial abnormalities in 52 school-age and adult patients with cleidocranial dysostosis. J Craniofac Genet Devel Biol. 13: 98-108, 1993.

20. Kelley A: Hereditary absence of clavicle. J Heredity 20: 353-355, 1929.

21. Kobayashi S, Uchida K, Baba H et al: Atlantoaxial subluxation-induced myelopathy in cleidocranial dysplasia. J Neurosurg Spine 7: 243-247, 2007.

22. Koch PE, Hammer WB: Cleidocranial dysostosis: review of the literature and report of a case. J Oral Surg 36: 39-42, 1978.

23. Langmead F: A family showing cleidocranial dysotosis. Proc Roy Soc Med: Sect Dis Child: 1-7: 1916-17

24. Lee B, Thirunavukkarasu K, Zhou L et al: Missense mutations abolishing DNA binding of the osteoblast-specific transcription factor OSF2/CBFA1 in cleidocranial dysplasia. Nat Genet 16: 307-310, 1997.

25. Marie P, Sainton R: Observation d'hydrocephalie heredtaire (per et fils) par vice de development du crane et du cerveau. Bull Mem Soc Med Hop Paris 14: 706-712, 1897.

26. Martin S: Sur un deplacement natural de la clavicule. J Med Chir Pharmacol 23: 456-460, 1765.

27. McBride ED: Congenital deficiency of the clavicle: hereditary cleidocranial dysostosis. J Bone Joint Surg 9A: 542-552, 1927.

28. McCurdy IJ, Baer RW: Hereditary cleidocranial dysostosis. JAMA. 81: 9-11, 1923

29. McGuire TP, Gomes PP, Lam DK et al: Cranioplasty for midline metopic suture defects in adults with cleidocranial dysplasia. Oral Surg, Oral Med Oral Pathol Oral Radiol Endod 103: 175-179, 2007.

30. Medhi N, Sahair B, Handique SK: Images: Progressive paraparesis with cleidocranial dysostosis. J Radiol Imaging 9: 141-143, 1999.

31. Mendoza-Londono R, Lee B: Cleidocranial dysplasia. Gene Review 1-18, 2006.

32. Morand GM Observations anatomique. Histoire de l'Academie Royale des Sciences Paris 4: 47, 1766.

33. Mundlos S: Cleidocranial dysplasia: clinical and molecular genetics. J Med Genet 36:177-182, 1999.

34. Mundlos S, Otto F, Mundlos C et al: Mutations involving the transcription factor cbfa1 cause cleidocranial dysplasia. Cell 89: 773-779, 1997.

35. Otto F, Thornell AP, Crompton T et al: CBFA1, a candidate gene for cleidocranial dysplasia syndrome is essential for osteoblast differentiation and bone development. Cell 89: 765-771, 1997.

36. Piatt JH Jr, Lagrange A: Congenital defects of the scalp and skull. In Principles and Practice of Pediatric Neurosurgery. Albright AL, Pollack IF, Adelson PD editors. New York, Thieme,1999, 209-218.

37. Rushton MA: An anomaly of cementum in cleidocranial dysostosis. Br Dent J 100: 81, 1956.

38. Scheuthauer G: Kombination rudimentarer Schulselbeine mit anomallien de schädels bei erwachsen menschen. Allg Wien Med Zig 16: 293-295, 1871.

39. Shaikh R, Shusterman S: Delayed dental maturation in cleidocranial dysplasia. ASDC J Dent Child 5: 325-329, 1998.

40. Shen Z, Zou CC, Yang RW, Zhao ZY: Cleidocranial dysplasia: report of 3 cases and literature review. Clin Pediat 48: 194-198, 2009.

41. Silva C, DiRienzo S, Serman N: Cleidocranial dysostosis: a case report. Col Dent Rev 2: 26-34, 1997.

42. Smith NH: A histologic study of cementum in a case of cleidocranial dysostosis. Oral Surg 25: 470-478, 1968.

43. Soule AB Jr.: Mutational dysostosis (cleidocranial dysostosis). J Bone Joint Surg. 28: 81-102, 1946.

44. Stein GS, Lian JB, van Wijnen AJ et al: RUNX2 control of organization, assembly and activity of the regulatory machinery for ske letal gene expression. Oncogene 23: 4325-4329, 2004.

45. Suba Z, Balaton G, Szabolcs G-G, et al: Cleidocranial dysplasia: diagnostic criteria and combined treatment. J Cranfac Surg. 16: 1122-1125, 2005

46. Tang S, Xu Q, Xu X et al: A novel RUINX2 missense mutation predicted to disrupt DNA binding causes cleidocranial dysplasia in a large Chinese family with hyperplastic nails. BMC Medical Genet 8: 82-88, 2007

47. Terry RJ: Rudimentary clavicles and other abnormalities of the skeleton of a white woman. J Anat Physiol 33: 13-422, 1899.

48. Thomsen G, Guttadauro M: Cleidocranial dysostosis associated with sclerosis and bone fragility. Acta Radiol 37: 559-567, 1952.

49. Tokuc G, Boran P, Boran BO: Cleidiocranial dysplasia in a mother and her daughter within the scope of neurosurgery. J Neurosurg 104: 290-292, 2006.

50. Vari R, Puca A, Meglio M: Cleidocranial dysplasia and syringomyelia-case report J Neuro Surg Sci 40: 125-128, 1996.

51. Visosky AM, Johnson J, Bingea B et al: Otolaryngological manifestations of cleidocranial dysplasia concentration on audiological findings. Laryngoscope 113: 1508-1514, 2003.

52. Wee LK, Pillay VK: Hereditary cleido-cranial dysostosis with a note on the anomaly of cementum. Singapore Med J. 5: 3-9, 1964.

53. Zheng Q, Sebald E, Zhou G et al: Dysregulation of chondrogenesis in human cleidocranial dysplasia. Am J Hum Genet. 77: 305-312, 2005.

54. Zhou G, Chen Y, Zhou L et al: CBFA1 mutation analysis and functional correlation with phenotypic variability in cleidocranial dysplasia. Hum Mol Gen 8: 2311-2316, 1999.

CHAPTER 8

Metaphyseal Dysplasias of Jansen and Schmid

Introduction:

In 1934, Murk Jansen a Dutch orthopaedic surgeon first described a rarely encountered autosomal dominant syndrome consisting of dwarfism, short and bowed lower extremities and rachitiform alterations (19). The syndrome has been found to be associated with hypercalcemia and may be related to genetic parathyroid disorders. In 1949, Franz Schmid a German physician described a similar syndrome but with normal calcium. In addition to the short bowed legs and dwarfism, the patients have limitation of finger extension and an extraordinary collagen disorder (37). Since then there have been numerous reports describing the similarities and differences between Type Jansen and Type Schmid metaphyseal dysostoses including some interesting biologic studies. It should also be noted that the McKusick syndrome resembles these disorders (25,26). It is known as cartilage hair dysplasia and is characterized by short stature, very thin colorless hair, bone deformities and an array of systemic complications (24).

Nomenclature and historical data:

The synonyms for both the Jansen and Schmid syndromes include Murk Jansen syndrome, Franz Schmid syndrome, dysostosis enchondralis, hereditary metaphyseal dysostosis, dysotosis metaphysaria, metaphyseal dysotosis, osteochondrodysplasia metaphysaria and osteochondritis subepiphysaria.

As indicated in the Introduction, Murk Jansen first reported the rare autosomal dominant dysplastic system in 1934 (19). His findings showed that the disorder is usually recognized by two years of age. It is characterized by short stature, marked shortening and equinovarus of the extremities and swelling of the joints. The children had normal mentation. The epiphyseal sites were enlarged and the metaphyseal areas calcified on X-ray. The description was further expanded by Cameron and Young in 1954 (3), Lenk in 1956 (23) and Evans and Caffey in 1958 (12). The findings for Jansen's disorder were further described by Gram and coworkers in 1959 (14) and by

Daeschner in 1960 (8). In 1959, DeHaas and coworkers (9) performed a late followup on Jansen's original patient and the findings supported the original suggestion of a calcified metaphyseal type of disorder, a position supported by Stickler and coworkers in 1962 (40). Further descriptions of the clinical presentations were provided by Rosenbloom and Smith in 1965 (32) and the X-ray changes described by Ozonoff in 1969 (30). Gordon and coworkers expanded the description of Jansen in a review of several patients in 1976 (13). It was Cooper and Ponseti in 1973 (6,7) who first defined the ultrastructural defect in the epiphyseal plates of affected patients and Holthausen and coworkers in 1975 (17), who reported changes in the calvarium on X-ray.

In 1949, Franz Schmid (37) described a similar disorder of somewhat milder nature and later onset. The patients however were short of stature and presented with more extensive bowing of the limbs with metaphyseal enlargements. The patients had a high frequency of hand deformities. The Schmid type was separated from the Jansen by a clearer description of the clinical findings by Weil in 1957 (44). Dent and Normand in 1964 (10) further identified the Schmid type of distinctive clinical changes. Arroyo-Scotoliff in 1973 (1), Koslowski in 1975 (20) and Spranger in 1977 (39) compared the clinical characteristics of the Schmid and Jansen types and also included some other genetic disorders with short stature and epiphyseal abnormalities.

Biologic characteristics for Jansen and Schmid disorders:

Patients with Jansen type of metaphyseal chondrodysplasia have severe agonist-independent hypercalcemia (4,21,31,34-36). The disorder appears to be caused by error in chromosome 6q21-q22.3 which results in mutations in PTHR1, a member of a distinct family of G protein-coupled receptors (34,35,45). The material is abundantly expressed in kidney and bone where it normally mediates the PTH-dependent regulation of calcium and phosphorus. It is also found in the growth plate where it mediates the pTHrP-dependent regulation of chondrocyte growth and differentiation (34-36,45). Four different mutations in the gene encoding the PTH/ PTHrP receptor (PTHR1) have been identified. Three of the PTHR1 mutations appear in severe form of Jansen's disease. These involve codon 223, codon 410 and codon 458 (45). Codon 410 alone has also been identified in a milder form of the disease and patients have less growth or joint problems. These mutant PtHR1s cause agonist-independent cyclic adenosine monophosphate (c-AMP) accumulation which affect the patient and the skeleton (35,36). The problem of such severe parathyroid abnormalities in these children not only result in short stature, but cause extensive damage and calcium deposition in the metaphyses and sometimes severe malformation of the body's bony structures (31,35).

Schmid's disease has no problem with parathyroid hormone or with calcium metabolism. Instead the difficulty appears to be a genetic error in the production of type X collagen (42). The chromosome 6q21-q22 produces a gene COL10A1, which is responsible for the production of Type X collagen (41). Type X is a short chain collagen expressed specifically by hypertrophic chondrocytes in the endochondral growth plate. Type X is also activated in fracture repair and in osteoarthritis. Mutations in COL10A1 cause errors in the production of type X collagen and result in damage to the bones and cartilage which result in metaphyseal chondrodysplasia type Schmid (41,42). The errors result in a similar pattern to that observed in patients with Jansen type disease, but the epiphyseal regions are abnormal and resemble rickets, and the metaphyseal regions of the bone are less prominent and not calcified (22,28)). The bones are somewhat less structurally disordered.

Clinical findings in patients with Jansen and Schmid disorders:

Type Jansen disease is a rarely encountered autosomal dominant disorder. The disease usually becomes manifest by one to two years of life (1,4,5,15,20,31,32,38). The children are short in height and mentally normal. They show severe abnormalities in the metaphyses, more prevalent in the lower extremities (1,4,15,20,31). The children have marked bowing deformities of the tibiae and femora (1,4,15,20,25,31,38). The gait is waddling and spine is deformed usually with anterior curvature. The children have a semi-squatting stance with flexion of the knees and hips (31,38). The patients have a bell-shaped thorax with widened costo-chondral junctions. Thoracic dysplasia can produce respiratory impairment and can be life-threatening (1,2,16,29,38,39). The fingers are short and clubbed. The facial appearance shows a receding chin, micrognathia, prominent eyes, a high arched palate and hypertelorism (1,17,31,38). Cranial problems can result in early deafness (17). From earliest onset the children have a sometimes profound hypercalcemia, hypophosphatasia and increased serum alkaline phosphatase (34-36).

Schmid type metaphyseal chondrodysplasia has similar clinical findings to those reported for Jansen type but are usually considerably milder. They appear somewhat later often after the age of two and principally affect the lower extremities (31,43). The findings closely resemble those of Vitamin D resistant rickets with remarkable changes in the epiphyseal plates on X-ray (22,28). The limbs are bowed and there is often a marked coxa vara (22), The knees also show a rather severe varus deformity and the children sometimes have difficulty walking. There is less effect on the spine and the face and dental problems are less abnormal so that the children have normal facies in contrast with the changes seen in those with Jansen type disorder (28). Cupping and flaring of the ribs is a common finding. The children are normal mentally and aside from their short stature and bowed extremities seem to function considerably better than patients with Jansen disorder. The hands are often involved and the children sometimes have difficulty extending digits (11). The calcium and phosphorus are normal and despite the ricket-like findings in the epiphyses, Vitamin D and calcium have no effect on their problem and should not be utilized (22,28).

Of some interest is the finding of changes very similar to Schmid type of metaphyseal dysplasia in Alaskan Malamute dogs (33). The animals are dwarfed, have bowed extremities and rachitic-like findings in the long bones.

The imaging studies for the two disorders are similar, particularly in terms of the bowing of the extremities (1,4,20,22,27,30,31,38,39,43). Patients with Jansen disorder show extensive metaphyseal enlargement and calcification (1,13,15,20,27). In addition the varus deformities of the bones are very prominent and sometimes quite severe. Ricket-like epiphyseal appearance but more normal bone structure is seen in Schmid disorder (22,31,38,43). The calvarium and facial bones are more often affected in Jansen type disease and sclerosis of the skull base is characteristic (4,5,17,27,31).

Treatment for children with Jansen and Schmid dysostoses:

Basically there are no therapeutic measures which alter the disease states for either of these entities. There are no treatment protocols to alter the genetic errors in the parathyroid system for children with Jansen type disease and Vitamin D and calcium are not only ineffective in treating the changes in patients with Schmid disorder, but can be dangerous. Bracing protocols may be helpful for children with severe deformities and osteotomies may have a role in improving function but are generally unnecessary, especially for type Schmid who are less affected (20,22). As they age, thoracic and cardiac disease may be a problem for patients with type Jansen disorder (29,31). Both types of patients may require some spine bracing or even surgery particularly for bowing deformities (16,18).

Discussion:

Metaphyseal chondrodysplasias (or dysostoses) are rare autosomal dominant genetic diseases . . . and as indicated in this chapter, come in two major forms. The first of these, Jansen type is caused by a very unusual metabolic abnormality affecting the parathyroid hormone genetic production system. The result is short stature, bone abnormalities, calcification in the metaphyses and spinal, calvarial abnormalities. The children also have biochemical problems affecting serum calcium and phosphorus. The second form, Schmid type is caused by a genetic error in the synthesis of collagen type X, which causes changes in the epiphyseal cartilage resembling rickets, but cannot respond to treatment with vitamin D. In addition, these patients are also of short stature, have bone abnormalities, spinal difficulties and some hand problems. They are somewhat less affected than the Jansen type patients but also have spinal and thoracic abnormalities.

Despite extensive clinical, imaging and biochemical studies on these patients we have no real reasonable approach to improving the lives of these children. In fact, despite their physical abnormalities and bizarre bony structure, they are mentally competent and for the most part live on into adulthood and sometimes beyond. They are less than competent physically but manage to maintain some degree of life participation.

It would be advantageous and indeed a great accomplishment if we were able to alter the genetic errors either in the parathyroid gland system or in the synthesis of collagen X and provide the children with an opportunity to grow normally and enjoy a productive life. The likelihood of that occurring is small chiefly because the disorders are so rare. Nevertheless, it seems reasonable to try to help these children. That would be a blessing!

References:

1. Arroyo-Scotoliff H: Metaphyseal dysostosis, Jansen type. J Bone Joint Surg 55A: 623-629, 1973.
2. Bethem D, Winter RB, Lutter L et al: Spinal disorders of dwarfism: review of the literature and report of eighty cases. J Bone Joint Surg 63A: 1412-1425, 1981.

3. Cameron JAP, Young WB, Sissons HA: Metaphyseal dysostosis. Report of a case. J Bone Joint Surg 36B: 622-629,1954.

4. Campbell JB, Kozlowski K, Lejam T, Sulko J: Jansen type of spondylometaphyseal dysplasia. Skel Radiol 29: 239-242, 2000

5. Charrow J, Poznanski AK: The Jansen type of metaphyseal chondrodysplasia: confirmation of dominant inheritance and review of radiographic manifestations in the newborn and the adult. Am J Med Genet. 18: 321-327, 1984.

6. Cooper RR, Ponseti IV: Metaphyseal dysostosis: description of an ultrastructural defect in the epiphyseal plate chondrocytes. Case report. J Bone Joint Surg. 55A: 485-495, 1973.

7. Cooper RR, Pedrini-Mille A, Ponseti IV: Metaphyseal dysostosis: a rough surface endoplasmic recticulum storage defect. Lab Invest 28: 119-125, 1973.

8. Daeschner CW, Singleton EB, Hill LL, Dodge WF: Metaphyseal dysostosis. J Pediatr 57: 844-854, 1960

9. De Haas WHD, De Boer W, Griffioen F: Metaphyseal dysostosis: a late followup on the first reported case. J Bone Joint Surg 51B: 290-299, 1969.

10. Dent CE, Normand ICS: Metaphyseal dysotosis, type Schmid. Arch Dis Childh 39: 444-454, 1964.

11. Elliott AM, Fiedl FM, Rimoin DL, Lachman RS: Hand involvement in Schmid metaphyseal chondrodysplasia. Am J Med Genet 132A: 191-193, 2005

12. Evans R, Caffey J: Metaphyseal dysostosis resembling Vitamin-D refractory rickets. Am J Dis Child 95: 640-648, 1958.

13. Gordon SL, Varano LA, Alandate A, Maisels MJ: Jansen's metaphyseal dysotosis. Pediatrics 58: 556-560,1976.

14. Gram PB, Fleming JL, Frame B, Fine G: Metaphyseal chondrodysplasia of Jansen. J Bone Joint Suig 41A: 951-959, 1959.

15. Jaffe HL: Metaphyseal dysostosis. In Metabolic, Inflammatory and Degenerative Diseases of Bones and Joints. Philadelphia, Lea and Febiger, 1976, 222-226.

16. Herring JA: The spinal disorders in diastrophic dysplasia. J bone Joint Surg. 60: 177-182, 1978.

17. Holthausen W, Holt JF, Stoeckenius M: The skull in metaphyseal chondrodysplasia type Jansen. Pediatr Radiol 3: 137-144, 1975.

18. Jalanko T, Remes V, Peltonen J et al. Treatment of spinal deformities in patients with diastrophic dysplasia. Spine 34: 2151-2157, 2009.

19. Jansen M Über atypisiche chondrodystophie (achondroplasie) und über eine noch nicht beschriebene angeborene wachstumsstörung des knochensystems: metaphysäre dystososis. Zschr Orthop Chir 61: 253-286, 1934.

20. Koslowski K: Metaphyseal and spondylometaphyseal chondrodysplasias. Clin Orthop 114: 83-93, 1976.

21. Kruse K, Schütz C: Calcium metabolism in the Jansen type of metaphyseal dysplasia. Eur J Pediatr 152: 912-915, 1993.

22. Lachman RS, Rimoin DL, Spranger J: Metaphyseal chondrodysplasia Schmid type: clinical and radiographic delineation with a review of the literature. Pediatr Radiol 18: 93-102, 1988.

23. Lenk R: Hereditary metaphyseal dysostosis. Am J Roentgenol 76: 569-575, 1956.

24. Mankin HJ: Cartilage hair hypoplasia. In Pathophysiology of Orthopaedic Diseases, Volume II. Rosemont, IL, American Academy of Orthopaedic Surgeons, 2009, 139-144.

25. McKusick VA: Metaphyseal dysostosis and thin hair: A "new recessively inherited syndrome. Lancet 1: 832-833, 1964.

26. McKusick VA, Eldridge R, Hostetler JA et al: Dwarfism in the Amish II: Cartilage hair dysplasia. Bull Johns Hopkins Hosp 116: 285-326, 1965.

27. Miller SM Paul W: Roentgen observations in familial metaphyseal dysostosis. Radiology 83: 665-673, 1964.

28. Nishimura G, Manabe N, Kosaki K et al: Spondylar dysplasia in type X collagenopathy. Pediatr Radiol 31: 76-80, 2001.

29. Nishioka K, Hiramatsu K: Systemic disease in bony thorax. J Japan Radiol Soc. 31: 1192-1197, 1972.

30. Ozonoff MB: Metaphyseal dysostosis of Jansen. Radiol 9: 1047-50, 1969.

31. Rimoin DL, Lachman RS: Genetic disorders of the osseous skeleton. In McKusick's Heritable Disorders of Connective Tissue. Fifth Edition, Editor: Peter Beighton. St. Louis, Mosby, 1993, 627-634,

32. Rosenbloom AL, Smith DW: The natural history of metaphyseal dysostosis. J Pediatr 66: 857-868, 1965.

33. Sande RD, Alexander JE, Spencer GR et al: Dwarfism in Alaskan malamutes: a disease resembling metaphyseal dysplasia in human beings. Am J Pathol 106: 224-236, 1982.

34. Schipani E, Jensen G, Pincus J et al: Constitutive activation of the cAMP signaling pathway by PTH/PTHrP receptors mutated at the two loci for Jansen's metaphyseal chondrodysplasia. Mol Endocrinol 11: 851-858, 1997.

35. Schipani E, Langman C, Hunzelman J et al: A novel parathyroid hormone (PTH)/ PTH-related peptide receptor mutation in Jansen's metaphyseal chondrodysplaisa. J Clin Endocrinol Metab. 84: 3052-3057, 1999.

36. Schipani E, Langman CB, Parfitt AM et al: Constitutively activated receptors for parathyroid hormone and parathyroid related peptidc in Jansen's metaphyseal chondrodysplasia. N Eng J Med 335: 708-714, 1996.

37. Schmid F: Beitrag zur dysotosis enchondralis metaepiphysaria. Mschr Kinderheilk 97: 393-397, 1949.

38. Silverthorn KG, Houston CS, Duncan BP: Murk Jansen's metaphyseal chondrodysplasia with long term followup. Pediatr Radiol 17:119-123, 1983.

39. Spranger JW: Metaphyseal chondrodysplasia. Postgrad Med J 53: 480-486, 1977.

40. Stickler G, Maher FT, Hunt JC et al: Familial bone disease resembling rickets (hereditary metaphyseal dysostosis). Pediatrics 29: 996-1004, 1962.

41. Wallis GA, Rash B, Sykes B et al: Mutations within the gene encoding the ⊠1(x) chain of type X collagen (COL10A1) cause metaphyseal chondrodysplasia type Schmid but not several other forms of metaphyseal chondrodysplasia. J Med Genet 33; 450-457, 1996.

42. Warman MI, Abbott M, Apte SS et al: A type X collagen mutation causes Schmid metaphyseal chondrodysplasia. Nature Genet 6: 79-82, 1993.

43. Wasylenko MJ, Wedge JH, Houston CS: Metaphyseal chondrodysplasia, Schmid type. J Bone Joint Surg 62A: 660-663, 1980.

44. Weil S: Die metaphysaren dysostosen. Zeit Orthop 89: 1-16, 1957.

45. Weir EC, Philbrick WM, Amling M: Overexpression of parathyroid hormone-related peptide in chondrocytes causes chondrodysplasia and delayed endochondral bone formation. Proc Natl Acad Sci USA 93: 10240-10245, 1996.

CHAPTER 9

Menkes Kinky Hair Disease: A Strange Genetic Entity

In 1962, John Menkes, a pediatrician at Columbia University in New York City reported 5 male infants who were affected by a distinctive syndrome of neurologic degeneration, peculiar hair structure, failure to thrive and an array of additional cardiac, vascular and skeletal problems. The children were seriously impaired and all died by 3-5 years of age. In 1972, David Danks further defined the problem and related it to a deficiency of copper and serum ceruloplasmin. Since then there have been extensive studies of this X-linked autosomal recessive disorder in an attempt to further define not only the reasons for reduction in copper in the tissues, but the relationship to other disorders with similar characteristics. Since the 1990's several investigators have attempted treatment protocols for newborn or very young children with forms of medication containing copper but with only modest success and only in the very young.

Nomenclature and history:

Menkes kinky hair disease is also known as MKHD, copper transport disorder, occipital horn syndrome, kinky hair disease, Menkes' syndrome and trichopoliodystrophy. A milder form of the same disorder is known as occipital horn syndrome (OHS). In addition, a clinical entity known as Wilson's disease is sometimes confused with this disease based on a similarity of copper problem dependency, but it is quite different in that the concentrations of copper are much increased, instead of diminished as in Menkes or occipital horn disease.

The history of Menkes kinky hair disease actually goes back to 1937 (4) when two Australian veterinarians, Bennetts and Chapman described a critical role of copper in mammalian neural development in ataxic lambs. The lambs had brittle dark colored wool, were chronically ill and died at an early age. The dependence on copper was related to the fact that the animal's mothers were thought to have grazed in copper deficient pastures (4).

In 1962, John Menkes, (30) a pediatrician in Columbia University of New York reported 5 male infants in one family, who were severely affected by a syndrome of neurogenic degeneration, peculiar hair, short stature and failure to thrive. They developed vascular abnormalities and epilepsy and died usually of cardiac disease in the first decade of life.

Aquilar et al (1) reported several cases in 1965 and O'Brien and Sampson (33) in 1966 suggested that the hair changes were particularly useful in identifying the disease although some patients were reported to have no hair either in their scalp or eyebrows. Billings and Degnan in 1971 (5) provided information about the clinical and pathologic characteristics of the disorder.

In 1972, David Danks and coworkers (9) of the Royal Children's Hospital in Melbourne, Australia related the Menkes disorder to the Australian sheep problems and postulated that the error lay within copper metabolism (8). They measured serum copper in seven patients and discovered that concentration was low, as was the amount of ceruloplasmin, an important copper enzyme (8). In 1972, French et al (12) and Hatak and coworkers (18) and in 1979, Grover et al (15) introduced the term "trichopoliodystrophy" for the disorder, thus defining it as affecting both hair and brain. Most clinicians and scientists, however prefer the name Menkes kinky hair disease (7,23,29). Since then a number of scientists have tried to define the biologic and genetic mechanism of the disease and perhaps equally importantly attempted to add copper to the affected child to allow them to grow and relieve them of their neurologic and other problems (7,21-23,25,29,32,41). Thus far the success rate is low but is particularly helpful if children can begin treatment shortly after birth (23).

Biologic and genetic characteristics of Menkes disease:

Menkes kinky hair disease is a rarely encountered X-linked recessive autosomal genetic disorder, which occurs almost entirely in male infants (3,7,19,21,29,30,41). There is no evidence of ethnic preponderance although Menkes' first five patients were of English-Irish origin (30) and Danks' patients were from Australia (8). The disease has been described in Japanese patients as well as from many European countries and Turkey and the overall incidence appears to be 1 in 70,000 live births (7,17,19,20,25). The striking biochemical findings are the low concentration of copper in plasma, liver and brain related principally to impaired intestinal absorption, reduced activities of copper dependent enzymes and paradoxical accumulation of copper in the duodenum, kidney spleen, pancreas and skeletal muscle (7,16,21,27,41). Children with the disease have a reduced catecholamine synthesis and have low serum levels of copper and ceruloplasmin and altered cerebrospinal fluid catechol levels (7,10,21,37). Specifically, there is a markedly diminished concentration of enzymes that require copper as a cofactor. These include dopamine-⊠-hydroxylase, cytochrome c oxidase and lysyl oxidase (7,10,27).

The genetic error appears to be located at chromosome X, band 13 (Xq13) (7,10,31, 35,44) and to result from a mutation in ATP7A and ATP7B, the two copper transport genes (7,10,11,23,35,36,42,44). Although there are many mutations identified in the ATP7A and B genes, the two major missense ones are A1362D and S637L (7,10,11,23, 27,35,42). The ATP7 genes are normally located on the trans-Golgi membrane of nearly all cells except hepatocytes and transport copper from the cytosol to the Golgi apparatus (7,10,11,35,40). Complete loss of function in the ATP7 genes result in markedly diminished copper transport in Menkes disease, while a less severe loss causes a milder allelic disorder, known as occipital horn syndrome (OHS) (7,11,21,27,31).

Clinical presentation in patients with Menkes kinky hair disease:

As indicated the disorder is an autosomal recessive X-linked genetic disorder so that the patients are almost always males. The pregnancy and delivery are usually normal although birth sometimes occurs slightly prematurely (7,23). Some of the children are found to have pectus excavatum and inguinal or umbilical hernias (7,17,23). The newborn children usually do not have hair abnormalities although there is sometimes less hair than is present with other children. Generally the children seem to be healthy and remain so for the first 4-6 weeks of life (7,17,21,23,25). It is not until the child reaches the age of 2-3 months that parents and physicians begin to notice problems. By the time the child reaches the age of 4-5 months he has developed progessive hypotonia, seizures, lethargy, failure to feed and the appearance of the diagnostic alterations in the structure of the hair. Untreated children generally die before the age of 9 (7,17,21,25).

Hair and facial features: The scalp hairs of children with Menkes kinky hair disease are short, sparse, coarse and twisted (7,15,25,46). The hair is often less abundant and even shorter on the sides and back of the head. The hairs are lightly pigmented and may display white, silver or gray colors and the eyebrows and eyelids are similarly affected (7,21,25,46). The face has pronounced jowls, sagging cheeks and relatively large ears (6,7,21). A "cupid's bow" upper lip is common (6). The palate is large and tooth eruption is delayed (6).

Body structure: Generally the patients are of short stature and sometimes have profound muscle weakness (7,17,23,25). One of the more common features is pectus excavatum and some mild to moderate scoliosis (7,20). The limbs are short and sometimes bowed. Joint motion is limited in patients with severe disease (20,25,45). The thumbs are held in an adducted position (7). Umbilical or inguinal hernias are often present. The skin is loose and redundant. X-rays of the skeleton show metaphyseal spurs on the femur and less often on the radius and humerus. Occasionally, excess calcification of long bones are noted (7,20,45). Wormian bony changes are seen in the calvarium (20).

Neurologic problems: Profound hypotonia with poor head control is invariably present.

Deep tendon reflexes are usually hyperactive (7,23,43). Visual fixation is impaired (38). Shortly after a year, the patients begin having seizures and epilepsy (2,7,28). Spasticity increases with advancing age (2,28). The children are quite severely mentally impaired which worsens with age (7,15,28).

Vascular and other abnormalities: Arterial studies show widespread elongation, tortuosity and lumen irregularities in all arteries (7,13,25). Arterial occlusion can develop and is often the cause of neural, aortic or cardiac problems and is the leading cause of death (7,13,21,25,43). Urologic abnormalities are known to occasionally occur (34,47) and even rarer are a series of sometimes fatal pulmonary problems (14).

It should be noted that there is a mild form of the disease but presenting with the same abnormality in copper genetics. It is known as Occipital Horn Syndrome and has some similarities to Menkes disease in terms of hair changes and some neurologic problems, but is often much milder (10,11,25). The disorder is distinguished and characterized by the presence of bony abnormalities in the form of projections from the metaphyses of the long bones and the calvarium (7,10,11). Patients with this disorder are not as impaired as those with Menkes and although their intellectual capacity is still lower than normal, they can communicate better and enjoy a more productive life. The patients live considerably longer than those with true Menkes and can sometimes be quite intellectually competent (7,11).

Diagnosis and treatment of Menkes disease:

Prenatal or neonatal diagnosis of the disorder is essential if one has any hope of helping the child to survive and be partially restored to a functional life (15-17,21,23,24). The traditional approach to identifying the disease is evaluating the hair changes and then measuring serum levels of copper and ceruloplasmin (7,16,17,21,23,24). The hair changes can be evaluated by histology examination, which will show a characteristic diagnostic feature known as "pili torti" (twisted hair) (46). These studies are not entirely reliable in detecting the disease in a new-born child. The actions of copper have recently been noted to increase the concentrations of dopamine, norepinephine, dihydroxyphenylacetic acid and dihydroxyphenylglycol. Measurements of these seem to be of considerable value, since all will be elevated in the serum of the affected child (10,23,24,26).

Once the disease is established administration of copper histidine at a dose of 250 micrograms twice daily by subcutaneous injection has been utilized and appears to improve the children's condition and survival rates (7,21-23,25,39,41,47).

There are however some additional approaches to management for the patients and their families which are essential.

1. Genetic counseling is an important element. The disease is transmitted to male children by asymptomatic mothers and hence, prenatal counseling and even intrauterine diagnosis can help as well (16,17,24).
2. Vaccination of children against a variety of diseases is important. This is especially true for influenza which can be life threatening to the patients (7.,23,25).
3. Prophylaxis against urinary tract infections is essential (34,47).
4. Consultation with neurologists, cardiologists, vascular surgeons, urologists and orthopaedists may be helpful in recognizing and treating some of the problems and especially in preventing development of serious consequences of the disease (7,23). Psychiatric consultations may be helpful for both the patients and their families. Physical and occupational therapy may improve the child's ability to care for himself.

Conclusions:

Menkes kinky hair disease has not only an unusual name but is an extraordinary disorder. The disease is caused by an autosomal genetic error in the X chromosome for the materials that utilize ingested copper to maintain the function of the brain, blood vessels, skeleton, urologic system and hair. The disease is not often apparent at birth but progresses fairly rapidly so that the male children are of short stature, mentally impaired and at risk for serious brain, cardiac or urologic problems by the age of 3 or 4 years. Death often occurs prior to age of 10 years. The disorder is further complicated by the occurrence of two other disorders of copper metabolism: a milder form known as occipital horn syndrome which presents with islands of increased production of bone, often in the metaphyses; and a much more complex disorder known as Wilson's disease, which is similar in genetic site but is the opposite in effect, notably that the patient has too much copper. These patients develop liver disease, which is a serious threat to their lives.

The treatment for Menkes kinky hair syndrome is currently an early diagnosis (not always easy) and administration of copper histidine from birth on. If given early enough, it seems to be successful in reducing the neural and vascular problems. Ideally, a genetic treatment would possibly solve the problem and there have been some successes with such protocols for animals. Thus far, however, this approach is only beginning to be of some potential value. If it can be done, it would solve the problem and not only would the children eliminate their neurologic, vascular, urologic and skeletal curses, but would almost surely grow normal hair! What a joy that would be!

References:

1. Aguilar MJ, Chadwick DL, Okuyama K et al: Kinky hair disease: I. Clinical and pathological features. Pediatrics 36: 417-420, 1965.
2. Bahi-Buisson N, Kaminska A, Nabbout R et al: Epilepsy in Menkes disease: analysis of clinical stages. Epilepsia 47: 380-386, 2006.
3. Barzegar M, Fayyhazie A, Gasemie B, Shoja MAM: Menkes disease: report of two cases. Iran J Pediatr 17: 388-392, 2007.
4. Bennetts HW, Chapman FE: Copper deficiency in sheep in Western Australia: a preliminary account of the aetiology of enzootic ataxia of lamb and an anemia of ewes. Aust Vet J 13: 138-149, 1937.
5. Billings DM, Degnan M: Kinky hair disease: a new case and a review. Am J Dis Child 121: 447-449, 1971.
6. Brownstein JN, Primosch RE: Oral manifestations of Menkes' kinky hair syndrome. J Clin Pediatr Dent 25: 317-321, 2001.
7. Danks DM: Menkes syndrome (kinky hair syndrome): In McKuskick's Heritable Disorders of Connective Tissue. Beighton, P, Ed. St. Louis, Mosby, 1993, 534-539.
8. Danks DM, Cartwright E, Stevens BJ et al: Menkes' kinky hair disease: further definition of the defect in copper transport. Science 179:1140-1142, 1973.
9. Danks DM, Stevens BJ, Campbell PE et al: Menkes' kinky hair syndrome. Lancet 1: 1100-1102, 1972.
10. De Bie P, Muller P, Wijmenga C, Klomp LWJ: Molecular pathogenesis of Wilson and Menkes disease: correlation of mutations with molecular defects and disease phenotypes. J Med Genet 44: 673-688, 2007.
11. Donsante A, Tang J, Godwin SC et al: Differences in ATP7A gene expression underlie intrafamilial variability in Menkes disease/occipital horn syndrome. J Med Genet. 4: 492-497, 2007.
12. French JH, Sherard ES, Lubell H et al: Trichopoliodystrophy 1. Report of a case and biochemical studies. Arch Neurol 26: 229-244, 1972.

13. Godwin SC, Shawker T, Chang B, Kaler SG: Brachial artery aneurysms in Menkes disease. J Pediatr 149: 412-415, 2006.

14. Grange DK, Kaler SG, Albers GM et al: Severe bilateral panlobar emphysema and pulmonary arterial hypoplasia: unusual manifestations of Menkes disease. 139: 151-155, 2005.

15. Grover WD, Johnson WC, Henkin RI: Clinical and biochemical aspect of trichopoliodystrophy. Ann Neurol. 5: 65-71, 1979.

16. Gu YK, Kodama H, Satol E et al: Prenatal diagnosis of Menkes disease by genetic analysis and copper measurement. Brain Dev 24: 715-718, 2002.

17. Gu YK, Kodama H, Shiga K et al: A survey of Japanese patients with Menkes disease from 1990-2003: incidence and early signs before typical symptomatic onset, pointing the way to earlier diagnosis. J Inhert Metab Dis 28: 473-478, 2005.

18. Hatak NR, Hirano A, Poon TP et al: Trichopoliodystrophy I. Report a case and biochemical studies. Arch Neurol 26: 229-244, 1972.

19. Horn N, Morton NE: Genetic epidemiology of Menkes disease. Genetic Epidemiol 3: 225-230, 1986.

20. Ichihashi K, Yano S, Kobayashi S et al: Serial imaging of Menkes disease. Neuroradiology 32: 56-59, 1990.

21. Kaler SG: Diagnosis and therapy of Menkes syndrome, a genetic form of copper deficiency. Am J Clin Nutr 67 (suppl): 1029S-1034S, 1998.

22. Kaler SG, Buist NR, Holmes CS et al: Early copper therapy in Menkes disease patients with a novel splicing mutation. Ann Neurol 38: 921-928, 1995.

23. Kaler SG, Holmes CS, Goldstein DS et al: Neonatal diagnosis and treatment of Menkes Disease. N Engl J Med 358: 605-614, 2008.

24. Kaler SG, Tumer Z: Prenatal diagnosis of Menkes disease. Pren Diag 18: 287-289, 1998.

25. Kodama H, Murata Y, Hobayashi M: Clinical manifestations and treatment of Menkes disease and its variants. Pediatr Int 41: 23-429, 1999.

26. Kodama H, Sato E, Yanagawa Y et al: Biochemical indicator for evaluation of connective tissue abnormalities in Menkes' disease. J Pediatr 142:726-728, 2003.

27. Linder MC, Hazegh-Azam M: Copper biochemistry and molecular biology. Am J Clin Nutr 63: 797S-811S, 1996.

28. Madsen E, Gitlin JD: Copper and iron disorders of the brain. Ann Rev Neurosci 30: 317-337, 2007.

29. Menkes JH: Kinky hair disease: twenty five years later. Brain Dev 10: 77-79, 1998.

30. Menkes JH, Alter M, Steigleder GK et al: A sex-linked recessive disorder with retardation of growth, peculiar hair and focal cerebral and cerebellar degeneration. Pediatrics 29: 764-769, 1962.

31. Moller LB, Tumer Z, Lund C et al: Similar splice-site mutations of the ATP7A gene lead to different phenotypes: classical Menkes disease or occipital horn syndrome. Am J Hum Genet 66: 1211-220, 2000.

32. Munakata M, Sakamoto O, Kitamure T et al: The effects of copper-histidine therapy on brain metabolism in patients with Menkes disease: a proton magnetic resonance spectroscopic study. Brain Dev 27: 297-300, 2005.

33. O'Brien JS, Sampson EL: Kinky hair disease II. Biochemical studies. J Neuropathol Exp Neurol 25: 523-530, 1966.

34. Oshio T, Hino M, Kirino A: Urologic abnormalities in Menkes' kinky hair disease: report of three cases. J Pediatr Surg 32: 782-784, 1997.

35. Payne AS, Gitlin JD: Functional expression of the Menkes disease protein reveals common biochemical mechanisms among the copper-transporting P-type ATPases. JBiol Chem 273: 3765-3770, 1998.

36. Petris MJ, Mercer JF, Camakaris J: The cell biology of the Menkes disease protein. Adv Exp Med Biol 448: 53-66, 1999.

37. Petris MJ, Stausak D, Mercer JF: The Menkes copper transporter is required for the activation of tyrosinase. Hum Mol Genet 9:2845-2851, 2000.

38. Seelenfreund MH, Gartner S, Vinger PF: The ocular pathology of Menkes' disease. Arch Ophthalmol 80: 718-720, 1968.

39. Sheela SR, Latha M, Liu P et al: Copper-replacement treatment for symptomatic Menkes disease: ethical considerations. Clin Genet 68: 278-283, 2005.

40. Steveson TC, Ciccotosto GD, Ma X-M, et al: Menkes protein contributes to the function of peptidylglycine α-amidating monooxyygenase. Endocrinology 144: 188-200, 2003.

41. Tumer Z, Horn N: Menkes disease: recent advances and new aspects. J Med Genet 34: 265-274, 1997.

42. Tumer Z, Moller LB, Horn N: Mutation spectrum of ATP7A, the gene defective in Menkes disease. Adv Exp Med Biol 448: 83-95, 1999.

43. Uno H, Arya S, Laxova R et al: Menkes' syndrome with vascular and adrenergic nerve abnormalities. Acta Pathol Lab Med 107: 286-289, 1983.

44. Vulpe C, Levinson B, Whitney S et al: Isolation of a candidate gene for Menkes disease and evidence that it encodes a copper transporting ATPase. Nature Genet 3: 7-13, 1993.

45. Wesenberg RL, Gwinn JL, Barnes GR: Radiological findings in the kinky hair syndrome. Radiology 93: 500-506, 1969.

46. Whiting DA: Structural abnormalities of the hair shaft. J Am Acad Dermatol. 16: 1-25, 1987.

47. Zaffanello M, Maffeis C, Fanos V et al: Urological complications and copper replacement therapy in childhood Menkes sydndrome. Acta Paediatr 95: 785-790, 2006.

CHAPTER 10

Pyle's disease: a rare and remarkably mild disorder

Pyle's disease was originally described in 1931 by Edwin Pyle, a physician in Waterbury Connecticut. He reported a five-year old child who presented with bilateral genu varum and on X-ray was found to have a marked increase in the width of the femoral shafts. The child had no other significant abnormalities. The finding was subsequently described as "Erlenmeyer Flasking", named after a glass container utilized for storing chemical fluids invented in 1861 by Emil Erlenmeyer. Pyle's disease, also known as metaphyseal dysplasia, is a very rare autosomal recessive disorder with males and females equally affected. The chief problem with the condition is related to the difficulty in distinguishing a number of other diseases which have similar Erlenmeyer flasking of the long bones. Although the changes in the distal femur in these diseases resemble those seen in children with Pyle's disease, the patients have other problems which are sometimes more extensive and disabling. One disorder, cranio-metaphyseal dysplasia, despite calvarial and facial and sometimes neurologic abnormalities is also sometimes labeled as Pyle's disease.

Nomenclature and historical data:

As indicated in the introductory statement, Pyle's disease has a somewhat confusing nomenclatural history. The disease was originally described in 1931 by Edwin Pyle (1891-1961), an orthopaedist in Waterbury, Connecticut (28) and it became known as Pyle's disease. In 1933, Cohn described another patient in Germany (3), so that the disorder acquired an additional name of Pyle-Cohn disease. Bakwin and Krider in 1937 (1) introduced the term "familial metaphyseal dysplasia" and in some settings the disorder became known as Bakwin-Krider disease. The term "cranio-metaphyseal dysplasia" which describes a clinically different entity is also known as Pyle's disease (11).

Soon after Pyle's original description, a series of papers described additional patients of this very rare disorder. In addition to Cohn's presentation in the German literature in 1933 (3) and the report of Bakwin and Krider in 1937 (1), Feld and coworkers (7) and Kimons (18) in 1954 described additional patients with what they termed "familial metaphyseal dysplasia". David and Palmer in 1958 (5) and Frew in 1961 (10) each added additional patients. Some of the patients had cranial and neurologic disorders and these were termed cranio-metaphyseal dysplasia and were cited as a form of Pyle's disease (37). It is important to note that Gorlin and coworkers in a signal study published in 1970 (11) proposed that the differences in presentation should separate the two disorders clinically as well as in nomenclature. A key finding in Pyle's disease is the presence of expanded and thinned distal femora and proximal tibiae termed as "Erlenmeyer Flasking" based on the resemblance to a type of conical chemical flask invented by Richard August Carl Emil Erlenmeyer in 1868 (19). The finding is not diagnostic for Pyle's disease since Erlenmeyer flasking is often a characteristic of other diseases including Gaucher disease, hypophosphastasia, osteopetrosis, Marfan's syndrome, pycknodysostosis and osteopathia striata (4,6,14,23). All of these however have both additional osseous and clinical characteristics which help to distinguish them from Pyle's disease.

Genetic and biologic characteristics of Pyle's disease:

Pyle's disease is a genetic disorder which is rarely encountered. The disease appears to be transmitted as autosomal recessive (29,33) and has no specific gender or ethnic frequency (2,7,9,33). Reports of the disease have come from India, Italy, England, Japan, Australia and South Africa (15,16,17,24,29,36,38) and all suggest that the disease is only rarely encountered in their countries. No genetic error has been identified although some of the other disorders with Erlenmeyer flasking have errors in collagen structure, elastic fibers, and for Gaucher disease in the production of glucosylceramide hydrolase as a result of an autosomal recessive inheritance of a gene error mapped at Iq21-q22 (6,22).

The principal biologic problems present in patients with standard Pyle's disease is the occurrence of Erlenmeyer flasking of the distal femur, proximal tibia and less commonly, several other sites in the upper and lower extremity (2,9,33). The cause of this finding appears to be related to a problem with the ring of Ranvier, a circumferential groove in the periphery of the epiphyseal cartilage which was described by Louis Antoine Ranvier in 1873 (31,32). The ring or as it is otherwise known, the groove of Ranvier consists of a circumferential indentation at the periphery of the growth plate which completely surrounds the bone at the termination of the epiphyseal cartilage (13,20). In an extraordinary experimental study performed by Shapiro, Holtrop and Glimcher in 1977 (35), and another by Oni in 1997 (26), it was determined that the Ranvier ring or groove contains a group of densely packed cells, which serve as progenitors which express osteocalcin (20,26). As a result the groove becomes the site of synthesis for osteoblasts, chondroblasts, fibroblasts and elastic fibers (12,20,35,39). The osteoblasts and chondroblasts form the cuff of bone which surrounds the end of the epiphysis and the beginning of the adjacent metaphysis (35). They communicate closely with the cambium layer of the periosteum of the shaft of the bone and add osseous tissue to that site. The fibroblasts extend from the groove to contact and attach to the periosteal fibrous layer (35). All three of these cellular elements serve to decrease and limit the width of the bone at the junction of the epiphysis with the metaphysis, a process known as "funnelization". As reported by Ferrara et al in 2003 (8), there is some evidence to support the concept that the process is activated by vascular endothelial growth factor (VEGF).

It is quite apparent that patients with Erlenmeyer flasking of the femoral shafts have some biologic disorder which interferes with the funnelization process and specifically in some way alters the activities in the Ranvier groove. In Gaucher disease, the presence of Gaucher cells containing large

63

amounts of glucosylceramide interferes with bone function and the patients develop osteoporosis and osteonecrosis (6,22,23). Under these circumstances, the Ranvier groove cannot carry on the critical functions and the bone becomes expanded, osteoporotic and at times, osteonecrotic (6,22). The other disorders which present with Erlenmeyer flasking have some similar interferences with osteosynthesis and production of fibroblasts and these lesions produce varying degrees of damage to the Ranvier funnelization process (2,14,15,20,22,23). The nature of the process which results in the Erlenmeyer flasking in patients with Pyle's disease is currently unknown but it is possible that an error in VEGF may be the cause (8).

Clinical characteristics of patients with Pyle's disease:

Pyle's disease otherwise known as metaphyseal dysplasia is a rare genetic entity, transmitted as an autosomal recessive disease and is evident early in life in affected children (2,5,9,16,18,29). The patients are of normal height and intelligence and have no complaints as a rule except when they fracture an extremity. Most of the patients have mild genu varum (2,5,9,16,23,27,34,35). The elbows sometimes lack full extension and widening of the lower femora and clavicles may be palpable (5,9,16,23,27,34,35). The patients have no facial dysmorphism or clinodactyly, polydactyly or syndactyly and there is usually no scoliosis or platyspondyly evident on physical examination (38). An occasional patient shows changes in the face consistent with the diagnosis of Crouzon's disease. These consist of mild hypertelorism, exophthalmos, strabismus, beaked nose, short upper lip, hypoplastic maxilla, and relative mandibular prognathism (25).

X-ray studies on patients with Pyle's disease disclose a frequently massive expansion of the metaphyses of the tubular bones (2,7,14,16,17,27,33,34,36). The cortices are thin. The calvarium and base of the skull are sometimes sclerotic and the vertebrae show minimal platyspondyly. The medial portions of the clavicles, the sternal ends of the ribs and the ischial and pubic bones are slightly widened (2,7,17,33,36).

One of the problems with the diagnosis of Pyle's disease is the occasional use of the term to include cranio-metaphyseal dysplasia, which in fact is a quite different entity. The disorder is more common and is transmitted as an autosomal dominant genetic error (11,15,30,33,37). The patients present with deafness, repeated episodes of cough and mouth breathing. They are sometimes deaf or partially blind and may have partial facial paralysis, dental problems and low intelligence (11,15,30,33,37). The patients have mild to moderate hypertelorism and marked broadening of the root of the nose. Imaging studies show frontal, paranasal and occipital hyperostosis or sclerosis. The changes in the long bones seen in Pyle's disease are absent or much less severe in cranio-metaphyseal dysplasia (11,15,30,33,37).

Treatment of patients with Pyle's disease:

Most patients with Pyle's disease require no treatment. They are clinically asymptomatic and with the exception of some changes in the shape of the lower extremities have no findings that require treatment. Unfortunately fractures are common and do require surgical management (2,5). Advising patients to avoid sports activities where they may be physically traumatized seems appropriate. In addition it is not unreasonable with patients with lower extremity deformities to consider corrective osteotomies (2,21,33). To date, no reports of treatment with bisphosphonates, Vitamin D or calcium have appeared but it is possible that these may help improve the bone density and reduce the risk of fractures.

Conclusions:

Pyle's disease or metaphyseal dysplasia is a rare and extraordinary entity which appears to present with only abnormalities in the shape of long bones. The patients are normal in all other respects and have no clinical problems aside from structural abnormalities and risk of fracture. Some of the problems with the disorder include distinguishing it from other genetic diseases which present with Erlenmeyer flask abnormalities of the long bones and other more problematical abnormalities. Seeing a child with Pyle's disease is uncommon in clinical practice but it seems essential to be certain that he or she is not suffering from cranio-metaphyseal dysplasia, Gaucher disease, pyknodysostosis, hypophosphatasia, osteopathia striata or osteopetrosis. If it is possible to define the genetic error for Pyle's disease this would make the physician's role much simpler and might make life easier for the patient and his or her family.

References:

1. Bakwin H, Krider A: Familial metaphyseal dysplasia. Am J Dis Child 53: 1521-1527, 1937.
2. Beighton P: Pyle disease (metaphyseal dysplasia). J Med Genet 24: 321-323, 1987.
3. Cohn M: Konstitutionelle hyperspongiosierung des skeletts mit paritellen riesenwuchs. Fortschr Roentgen 47: 293, 1933.
4. Culver GJ, Thumasathit C: Osseous changes of osteopathia striata and Pyle's disease occurring in a patient with an 11 year follow-up. A case report. Am J Roentgenol Radium Ther Nucl Med 116: 640-643, 1972.
5. David JEA, Palmer PES: Familial metaphyseal dysplasia. J Bone Joint Surg 40B 86-93, 1958.
6. Elstein D, Itzchaki M, Mankin HJ: Skeletal involvement in Gaucher disease. Baillieres Clin Haematol 10: 793-816, 1997.
7. Feld H, Switzer RA, Dexter MW, Langer EM: Familial metaphyseal dysplasia. Radiology 65: 206-221, 1954.
8. Ferrara N, Gerber HP, Lecouter J: The biology of VEGF and its receptors. Nat Med 9: 599-676, 2003.
9. Ferrari D, Magnani M, Donelli O: Pyle's disease: a description of two clinical cases and a review of the literature. Chir Organi Mov 90: 303-307, 2005.
10. Frew JFM: Familial metaphyseal dysplasia. J Bone Joint Surg. 43B: 188, 1961.
11. Gorlin RJ, Koazalkia MF, Spranger J: Pyle's disease (familial metaphyseal dysplasia): a presentation of two cases and argument for its separation from craniometaphyseal dysplasia. J Bone Joint Surg 52A: 347-354, 1970.
12. Gigante A, Greco F, Specchia N, Nori S: Distribution of elastic fiber types in the epiphyseal region. J Orthop Research. 14: 810-817, 2005.
13. Girault JA, Peles E: Development of nodes of Ranvier. Curr Opin Neurobiol 12: 476-485, 2002.

14. Greenspan A: Sclerosing bone dysplasias—a target-site approach. Skeletal Radiol. 20: 561-583, 1991.

15. Gupta D, Sharma OP, Chaudhary AK, Gupta SK: Cranio-metaphyseal dysplasia. Australas Radiol 37: 122-125, 1993.

16. Gupta N, Kabra M, Das CJ, Gupta AK: Pyle metaphyseal dysplasia. Indian Pediatr 45: 323-32, 2008.

17. Heselson NG, Raad MS, Hamersma H et al: The radiological manifestations of metaphyseal dysplasia (Pyle disease). Br J Radiol 52: 431-440, 1979.

18. Komins C: Familial metaphyseal dysplasia: Pyle's disease. Brit J Radiol. 27: 670-675, 1954.

19. Krätz O: Das portrait: Emil Erlenmeyer 1825-1909. Chemi in Unserer Zeit 6: 3-58, 1972.

20. Langenskiold A: Role of the ossification groove of Ranvier in normal and pathologic bone joints: a review. J Pediatr Orthop 18: 173-177, 1998.

21. Lindberg, EJ, Watts HG: Post-osteotomy healing in Pyle's disease. Clin Orthop 341: 215-217, 1997.

22. Mankin HJ, Rosenthal DI, Xavier R: Gaucher disease: new approaches to an ancient disease. J Bone Joint Surg. 83: 748-762, 2001.

23. Mankin HJ, Sims KB, Bove CM: Erlenmeyer flasking of a child's bones: a diagnostic puzzle. Am J Orthop 34: 393-395, 2005.

24. Narayananan VS, Ashok L, Mamatha GP et al: Pyle's disease: an incidental finding in a routine dental patient. Dento-Maxillo-Facial Radiol 35: 50-54, 2006.

25. Nardi P, Biagi P, Poggini M, Mara M: Craniofacial dysostosis (Crouzon's disease) associated with metaphyseal dysplasia (Pyle's disease) in the same subject. Panminerva Med 31: 192-197, 1989.

26. Oni OOA: Osteocalcin expression in the groove of Ranvier of the rabbit growth plate. Injury 28: 109-111, 1997.

27. Percin EF, Percin S, Koptagel E, Demirel H: A case with Pyle type metaphyseal dysplasia: clinical radiological and histological evaluation. Genet Couns 14: 387-393, 2003.

28. Pyle E: Case of an unusual bone development. J Bone Joint Surg. 13: 874-876, 1931.

29. Raad MS, Beighton P: Autosomal recessive inheritance of metaphyseal dysplasia (Pyle disease). Clin Gent 14: 251-256, 1978.

30. Ramseyer LT, Leonard JC, Stacy TM: Bone scan findings in craniometaphyseal dysplasia. Clin Nucl Med 18: 137-139, 1993.

31. Ranvier L: Quelques faits relatifs au developpement du tissue osseux. Comptes Rend Acad Sciences. 77: 1105-1109, 1873.

32. Ranvier L: Traite technique d'histologie. Ed 2. Paris, Savy 1889, 356-357.

33. Rimoin DL, Lachman RS: Pyle disease: In McKusick's Heritable Disorders of Connective Tissue 5th Edition, Beighton P, editor. St. Louis, Mosby,1993, 663-666.

34. Ross MW, Altman DH: Familial metaphyseal dysplasia, review of the clinical and radiologic feature of Pyle's disease. Clin Pediatr 6: 143-149, 1967.

35. Shapiro F, Holtrop ME, Glimcher MJ: Organization and cellular biology of the perichondrial ossification groove of ranvier: a morphologic study in rabbits. J Bone Joint Surg 59: 703-723, 1977.

36. Shibuya H, Suzuki S, Okuyama T, Yukawa Y: The radiological appearances of familial metaphyseal dysplasia. Clin Radiol 33 439-444, 1982.

37. Small PO, Wallis JJ: Pyle's disease or craniometaphyseal dysplasia tarda. Brit J Radiol 43: 811-813, 1970.

38. Turra S, Gigante C, Pavanini G, Bardi C: Spinal involvement in Pyle's disease. Pediatr Radiol. 30: 25-27, 2000.

39. Walter K, Tansek M, Tobias ES et al: COL2A1-related skeletal dysplasias with predominant metaphyseal involovement. A, J Med Genet 143A 161-167, 2007.

CHAPTER 11

Weill-Marchisani Syndrome

Introduction

The Weill-Marchisani syndrome (WMS) is a rare and very unusual genetic disorder characterized by proportionate short stature, stiff joints, brachydactyly, stubby hands and feet, malformed and malaligned teeth, broad skull, small shallow orbits and sometimes devastating ocular and cardiac abnormalities. Late in the course many patients have ossification in their epiphyses and intervertebral disks. In most patients and their families, the gene error appears to cause an autosomal recessive error leading to the production of an extracellular protein, ADAMTS10. Much less commonly the disease is autosomal dominant and the material present is FBN1. In both forms of the disease, the patients appear to be mentally normal and survive a long time but have a high likelihood of blindness in the early decades and cardiac disease later in the course.

Nomenclature and history for Weill-Marchisani Syndrome

Weill-Marchisani syndrome is also known as WMS, spherophakia-brachymorphism syndrome, brachymorphism and ectopia-lentis syndrome, dysmorpho-dystrophia mesodermalis congenita, dystrophia mesodermalis congenita, dystrophia mesodermalis hypoplastica, and inverted Marfan syndrome.

In 1932, GeorgesWeill, (26) a French ophthalmologist, reported 8 cases of "inverted" Marfan's syndrome in which the patients were of short stature, had ectopia lentis and "short, swollen fingers permitting only imperfect opening and closing" of the hands. In 1939, Oswald Marchesani (13) a professor of ophthalmology in Hamburg, gave the first definitive description based on evaluation of two consanguinitous families He clearly described the spherophakia and brachymophia. Seeleman in 1949 (24), described the bone abnormalities in considerable detail. In 1955 and 1956, Rosenthal and Kloepfer (10,21), further described the disorder and the genetic characteristics and Zabriskie and Reisman in 1958 (31) termed it the Marchisani syndrome. Saxena et al (22) in 1966 described in detail, an Indian girl with dwarfism and glaucoma. In 1969, Rennert (20) described the clinical characteristics of WMS and Scott (23) introduced the name Weill-Marchisani syndrome. In 1971, Victor McKusick (15) examined the family members described by Rosenthal and Kloepfer and confirmed the genetic nature of the diagnosis and its clinical characteristics.

Genetic and biologic characteristics of Weill-Marchesani syndrome

ADAMTS are a series of metalloproteases, which are involved in a number of body processes including connective tissue organization and turnover, blood coagulation, inflammation, arthritis, angiogenesis and cell migration (7,12,19,32). The type of genetic mutation of ADAMTS which appears to be the cause of symptoms in patients with autosomal recessive WMS is ADAMTS10, which arises from a gene-map locus of 19q13.3-p13.2 and interacts with fibrillin, a major constitutive element of extacellular microfibrils (4,5,. Fibrillin has widespread distribution and critical role in both elastic and non elastic connective tissue throughout the body (3,4.7,19). Another form of the disease, which is autosomal dominant has as a genetic mutation material, FBN1 which arises from the gene map location 15q21.1 (2,29). Both of these errors are similar in terms of their principal effect being in the eyes and both cause changes in the stature, hand function and some characteristics in the extremities. Both are equally distributed in male and females and although cases have been described from many foreign countries, there is no detectable ethnic frequency (2,11,19,14,23).

Both ADAMTS10 and FBN1 can be assessed biologically in the patient to establish the diagnosis (19). It should be noted that the ADAMTS errors are also believed to be related to the genesis of Ehlers-Danlos disease, von Willibrand's disease, rheumatoid arthritis, Marfan's disease and some other less common disorders (7,12,19,32).

Clinical presentation of patients with Weill-Marchisani syndrome

Although the genetic error for the autosomal recessive and autosomal dominant forms of Weill-Marchesani syndromes are different, the clinical presentation is essentially identical (14,15,23,27). The patients are of short stature, but remarkably they are essential normal in appearance despite an average height of 142 centimeters (56 inches or 4.6 feet) (14,15,23). They have brachydactyly (shortening of digits of the hand) but the symmetry and finger structure is normal. Joint stiffness is quite common and the patients may not be able to flex their fingers sufficiently to grasp objects (2,4,14,20,23). They may have difficulty extending their elbows or shoulders normally. The skull is broad and the orbital spaces are smaller than normal (1,14,21,22,27). Patients have a narrow palate and mild maxillary hypoplasia. The teeth may be malformed and malaligned (14,15). Occasionally calcification can be noted in the epiphyseal-mtaphyseal regions of the long bones (14).

Ophthalmologic problems sometimes dominate the patient's condition. The eyes show some of the most significant alterations in structure. Microspherophakia is a congenital, usually bilateral condition, in which the crystalline lens is smaller than normal and spherical in shape (11,14,21,23,27). It may give rise to lenticular myopia, subbluxation or glaucoma and is most often present in Weill-Marchesani syndrome and less commonly in Marfan's disease

(1,14,22,27,30). Many of the patients have ectopia lentis (dislocation of the lens as a result of disruption of the zonular fibers of the lens), which may cause marked visual disturbance (16,17,27). The disorder may be present at birth or occur even late in life in affected WMS patients. It should be noted that ectopia lentis also occurs in patient with Marfan's syndrome, homocystinuria, Ehlers Danlos syndrome and syphilis (16,17,18,25,27). Glaucoma and optic atrophy are common events in these patients and are often bilateral and cause blindness (1,22,30). Retinitis pigmentosa, a disorder associated with degeneration of the retinal pigment epithelial cells, has been described and is another cause of blindness (6,25).

Cardiac abnormalities are less common than the ocular problems. A prolonged QTc on ECG occurs frequently and mitral valve prolapse is one of the problems, which may cause serious impairment of cardiac function (9,14). Many of the patients become dyspneic and develop chest pain (14).

Treatment of patients with Weill-Marchesani syndrome

Many patients with WMS do not require treatment for their extremity problems. Although the hands, wrists, elbows and shoulders may be limited in functions, the patients have little pain and often find methods to overcome the difficulties encountered. The eye problems are sometimes severe and lead to blindness based on lenticular myopia, ectopia lentis, glaucoma and retinitis pigmentosa (1,6,16,22,25,30). Surgery is often necessary but sometimes is not successful in eliminating visual loss and blindness (1,8,18,28,30). Cardiac disorders can be serious and may cause severe impairment. Supportive medication, and sometimes surgery is essential (9).

Conclusions:

Weill-Marchesani syndrome is a very rare disorder and despite the alteration in hand and upper extremity function, orthopaedists are rarely involved in the care of these patients. In contrast, the opthalmologists are frequently required to help solve the threat to vision in association with microspherophakia, glaucoma and ectopia lentis. The patients are short in stature but generally well formed and are normal in intelligence. They may have some serious psychiatric issues related to their orthopaedic and ocular problems.

It would be a wonderful accomplishment to somehow alter the genetic error or interfere with the ADAMST10 or FBN1 materials which are the cause of the problems for the patient. Thus far, there does not seem to be active approaches to all of these issues, partly because the disease is so rare and the patients are often quite functional.

References:

1. Asaoka R, Kato M, Suami M et al: Chronic angle closure glaucoma secondary to frail zonular fibers and spherophakia. Act Ophtalmol Scand 81: 533-535, 2003.

2. Chung JL, Kim SW, Kim JH et al: A case of Weill-Marchesani syndrome with inversion of chromosome 15. Korean J Opthalmol 21: 255-260, 2007.

3. Clark IM, Parker AE: Metalloproteinases: their role in arthritis and potential as therapeutic targets. Expert Opin Ther Targets 7: 2122-2127, 2003.

4. Dagoneau N, Benoist-Lasselin C, Huber C et al: ADAMTS10 mutations in autosomal recessive Weill-Marchesani syndrome. Am J Hum Genet 74: 801-906, 2004.

5. Faivre L, Megarbane A, Alswaid A et al: Homozygosity mapping of a Well-Marchesani syndrome locus to chromosome 19p13.3-p13.2. Hum Genet 110: 366-370, 2002.

6. Jethani J, Mishra A, Shetty S, Vijayalakshmi P: Weill=Marchesani syndrome associated with retinitis pigmentosa. Ind J Opthalmalog 55: 142-143, 2006.

7. Jones GC: ADAMTS proteinases: potential therapeutic targets? Curr Pharm Biotechnol 7: 25-31, 2006.

8. Harasymowycz P, Wilson R: Surgical treatment of advanced chronic angle closure glaucoma in Weill-Marchesani syndrome. J Pediatr Ophthalmol Strabismus 41: 295-299, 2004.

9. Kojuri J, Reza Razeghinejad M, Aslani A: Cardiac findings in Weill-Marchesani syndrome. Am J Med Genet. 145A: 2062-2064, 2007.

10. Kloepfer HW, Rosenthal JW: Possible genetic carriers in spherophakia-brachymorphia syndrome. Am J Hum Genet 7:398-425, 1955

11. Kulkarni ML, Venkataramana V, Sureshkumar-Satishchandra C: Weill-Marchesani syndrome. Indian Pediatr. 32:923-926, 1995.

12. Levy GG, Hichols WC, Lian EC et al: Mutations in a member of the ADAMTS gene family cause thrombotic thrombocytopenic purpura. Nature 413: 488-494, 2001.

13. Marchesani O: Brachydaktylie und angeborene Kugellinse als systemerkrankung. Klin Mbl Augenheilk 103: 392-406, 1939.

14. Maumanee IH: The Weill-Marchesani Syndrome: In McKusick's Heritable Disorders of Connective Tissue 5th Edition. Editor: Peter Beighton. St. Louis, Mosby, 1993, 179-187.

15. McKusick VA. Heritable Disorders of connective Tissue Ed 4: St Louis Mosby-Year Book, 1972.

16. Nelson L: Ectopia lentis in childhood. J Pediatr Ophthalmol Strabismus. 45: 12-, 2008.

17. Nelson LB, Maumenee IH: Ectopia Lentis. Surv Ophthalmol 27: 143-160, 1982.

18. Omulecki W, Wicaynski M, Gerkowicz M: Management of bilateral ectopia lentis et pupillae syndrome. Ophthalmic Surg Lasers Imaging. 37: 68-71, 2006.

19. Porter S, Clark IM, Kevorkian l, Edwards DR: The ADAMTS metalloproteinases. Biochem J. 385:15-27, 2005.

20. Rennert RN: The Marchesani syndrome: A brief review. Am J Dis Child 117:703-705, 1969.

21. Rosenthal JW, Kloepfer HW: The spherophakia-brachymorphia syndrome. Arch Opthalmol 55: 28-35, 1956.

22. Saxena JN et al: Dwarfism with brachydactyly, spherophakia and glaucoma. Marchesani syndrome in an Indian girl. Indian Pediatr 3: 231, 1966.

23. Scott CI: Weill-Marchesani syndrome. In Bergsma D, Editor: Clinical Delineation of Birth Defects. II. Malformation Syndromes. New York, National Foundation, March of Dimes, 1969, 238-240.

24. Seeleman K: Brachydaktylie und angeboren Kugellinse. Z Kinderheilk. 67: 1-6, 1949.

25. Sato H, Wada Y, Abe T:et al: Retinitis pigmentosa associates with ectipia lentis. Arch Ophthalmol 120: 852-854, 2002.

26. Weill G: Ectopie du cristllins et malformations generales. Ann Ocul 169: 21-44, 1932.

27. Wentzloff JN, Kaldawy RM, Chen TC: Weill-Marchesani syndrome. J Pediatr Ophthalmol Strabismus. 43: 192-, 2006.

28. Willoughby CE, Wishart PK: Lensectomy in the management of glaucoma in spherophakia. J Cataract Refract Surg 28:1061-1064, 2002.

29. Wirtz MK, Samples JR, Kramer PL et al: Weill-Marchesani syndrome-possible linkage of the autosomal dominant form to 15q21.1 Am J Med Genet. 65: 68-75, 1996.

30. Wright KW, Chrousos GA: Weill-Marchesani syndrome with bilateral angle closure glaucoma. J Pediatr Ophthalmol Strabismus 22: 129-132, 1985.

31. Zabriskie J, Reisman M: Marchesani syndrome. J Pediatr 52: 159-169, 1958.

32. Zheng X, Chung D, Takayama TK et al: Structure of von Willebrand factor-cleaving protease (ADAMTS13), a metalloprotease involved in thrombotic thrombocytopenic purpura. J Biol Chem. 276: 41049-41063, 2001.

CHAPTER 12

Craniodiaphyseal Dysplasia: A Very Rare Syndrome

Introduction:

Craniodiaphyseal dysplasia was first described in 1949 by Halliday but not really designated as separate entity until the report by Joseph in 1958. The disease consists of two major features: Erlenmeyer flasking and increase in the width and density of the long bones with sparing of the epiphyses and metaphyses; and marked hyperdensity and deformity of the calvarium. The disorder is rare, occurs in young children and produces a profound alteration in the appearance of the head and face. The appearance of the child's face is so extraordinary that it was originally called leontiasis ossea and frequently results in epilepsy and a progressive loss of vision and hearing. Density increase and malformation in the long bones frequently causes deformity and sometimes fractures and multiple joint difficulties. The children have short stature, mental retardation, marked disability and die at an early age. It is important to note that this rare disease resembles some other somewhat more common genetic disorders of the skeletal system but has some major differences, principally in terms of the remarkable appearance of the face and skull.

Nomenclature and History:

Craniodiaphyseal dysplasia is otherwise known as craniotubular dysplasia, ectodermal dysplasia, frontometaphyseal dysplasia and leontiasis ossea. Of some concern and frequent comment is the resemblance of this disorder to Van Buchem's disorder and Camurati-Engelmann's disease, both of which have Erlenmeyer flasking and diaphyseal hyperdensity but have little abnormality affecting the calvarium.

The first description of the disorder was by Gemmel in 1935 (10) who described the facial abnormality as leontiasis ossea. Halliday in 1949 (14) decided that the disorder was a rare case of bone dystrophy. In 1958, Joseph, a French physician and his colleagues (17) reviewed the previously reported cases and decided that the entity was unique and that the diagnosis should be termed progressive craniodiaphyseal dysplasia. Stransky and coworkers believed the disorder was a form of juvenile Paget's disease (33). Gorlin et al in 1969 (11,12) cited a series of cases which had been previously described as leontiasis ossea (craniofacial deformity leading to a lion-like face). MacPherson in 1974 (21) described three new cases and pointed out the variable manifestations of the disorder and the similarity to other diaphyseal hyperostotic genetic diseases. Subsequent studies by Kaitila et al in 1975 (18) and Tucker and coworkers in 1976 (34) further confirmed the quite specific nature of the skull and face changes and the distinction from Van Buchem's disease and the much more frequently encountered Camurati-Engelmann's disease.

Biologic characteristics for craniodiaphyseal dysplasia:

Craniodiaphyseal dysplasia is a very rare disorder with probably less than 100 cases reported until now (5,6,16,19,22,29,32,34) The disease is believed to be autosomal recessive and has no gender or ethnic specificity. A possible source of the genetic error is believed to be a mutation in 17q11.2 (6). The disease is believed to occasionally share a genetic abnormality with Van Buchem's disease consisting of a mutation in low-density lipoprotein receptor-related protein 5 gene (LRP5) (4,20,35). There are no recorded cases which have abnormalities of TGFB1 and SOST which tend to exclude the diagnoses of Camurati-Engelmann disease and other disorders which present with increased bone density (1,20,21,23). Patients with craniodiaphyseal dysplasia have been found to have increased PTH and serum alkaline phosphatase levels, but no other abnormalities of thyroxine, phosphate, calcium or vitamin D (2,5,19,32). The histology and X-ray imaging do not support the diagnosis of juvenile Paget's disease (3,6,32).

Clinical presentation of craniodiaphyseal dysplasia:

The clinical features of craniodiaphyseal dysplasia is often noted in children by the age of two (5,6.19,32). The children are short statured and have mental retardation (5,6,19,32). Because of their facial problems, many never leave their homes. Death usually occurs by the age of 20 despite attempts to treat them with drugs and surgical corrections of some of their problems (5,6,19,32). As far as the bones are concerned, the diaphyses of the long bones show a marked increase in density (2,3,5,6,13,16,19,24). The bones are widened and the femur, tibia and humerus often show Erlenmeyer flasking (2,5,6,9). The clavicles and ribs are also increased in width and are more dense than normal (5,616). The striking feature however is the extraordinary deformity of the skull and facial bones (2,6,15,19,24). Frontal and paranasal bossing, paranasal sinus and mastoid obliteration, hypertelorism and dental malocclusion all cause severe deformity of the face (2,6,15,24,25,28,32). The clinical features are remarkable and patients with this disorder are described as having a "lionization" of the face (2,5,7.28,33) A famous individual named Roy L. Dennis (1961-1978) known as "Rocky" stimulated the development of a 1985 film drama name "Mask". His body was donated to UCLA Medical Center after his death and material is still available to observe his facial deformity. The cranio-facial-cervical abnormalities cause dental and nasal abnormalities, hearing loss, severe ocular problems leading to blindness, respiratory impairment, and neurologic deficits in the neck and upper extremities (2,6,15,24,28,27,32). Despite efforts to treat them, most patients die before the second decade.

It is essential to note that craniodiaphyseal dysplasia has some clinical and radiologic features which in some patients resemble Van Buchem's, Camurati-Engelmann and Ribbings diseases but none of the three have similar severe abnormalities in the face and calvarium (6,8,12,16,21,22,23,30,32). Furthermore the prognosis and mental status for patients with craniometaphyseal dysplasia is quite poor and to some extent is quite different from the other disorders (6,21,23,30).

Treatment for patients with craniodiaphyseal dysplasia:

With so few patients that are available for treatment trials, few protocols have been suggested. The patients and their families are difficult to deal with mostly because the children are so deformed and disabled. Mental retardation and facial deformity makes the patient difficult to treat (25). Respiratory support may be necessary and dental restructuring may sometimes be necessary and helpful (6,19,25). There is no real way to prevent or treat hearing loss or visual disturbances except for surgical restructuring of the calvarium, which is considered difficult (31). Cervical surgery is sometimes necessary as well particularly for damage to the cord or nerve roots (28), Peripheral bone problems are sometimes severe, particularly with fractures, deformities, malalignments and joint abnormalities (2,5,19,32). Surgery is sometimes necessary and may be helpful.

Calcitonin has been administered to some patients with the hope that the drug will diminish excessive bone in the calvarium and limbs but the number of cases treated thusfar has not provided a clear statement as to success (26). Bisphosphonates have also been tried but have not provided any change in the limb status (19). Without any clear definition of the genetic status of the disorder there have been no approaches to gene therapy. Patient and family psychologic support seems appropriate, but is unlikely to change aspects of the problem or the outcome.

References:

1. Balemans W, Ebeling M, Patel N et al: Increased bone density in sclerosteosis is due to the deficiency of a novel secreted protein (SOSO). Hum Mol Genet. 10: 537-543, 2001.
2. Bieganski T, Baranska D, Miastkowska I et al: A boy with severe craniodiaphyseal dysplasia and apparently normal mother. Am J Med Genet 143A: 2435-2443, 2007.
3. Bonucci E, Menichini G, Scarlo GB, Tomaccini D: Histologic, microadiographic and electron microscopic investigations of bone tissue in a case of craniodiaphyseal dysplasia. Virchows Arch Pathol Anat Histol 373: 167-175, 1977.
4. Boyden lM, Mao J, Belsky J et al: High bone density due to a mutation in LDL-receptor-related protein 5. N Eng J Med 16: 1515-1521, 2002.
5. Brueton LA, Winter RM: Craniodiaphyseal dysplasia. J Med Genet 27: 701-706, 1990.
6. Canepa G, Maroteaux P, Pietrogrande V: Craniodiaphyseal dysplasia: in Dysmorphic Syndromes and Constitutional Diseases of the Skeleton, Volume 1. Padova, Italy, Piccin 2001, 451-454.
7. De Souza O: Leontiasis ossea case reports. Porto Alegre Fac Med 13: 46-54, 1972.
8. Eastman JR, Bixler D: Generalized cortical hyperostosis (Van Buchem disease). Nosologic considerations. Radiology. 125: 297-304, 1977.

9. Faden MA, Krakow D, Ezgu F et al: The Erlenmeyer flask bone deformity in the skeletal dysplasias. Am J Med Gent 149A: 1334-1345, 2009.

10. Gemmel JH: Leontiasis ossea: a clinical and roentgenographic entity. Radiology 25: 723-729, 1935.

11. Gorlin RJ, Cohen MM: Frontometaphyseal dysplasia. A new syndrome Am J Dis Child 118:487-494, 1969.

12. Gorlin RJ, Spranger J, Koszalka MF: Genetic craniotubular bone dysplasias and hyperostoses: a critical analysis. Birth Defects V: 79-95, 1969.

13. Greenspan A: Sclerosing bone dysplasia: a target-site approach. Skeletal Radiol 20: 561-583, 1991.

14. Halliday J: A rare case of bone dystrophy. Br J Surg 37: 52-63, 1949.

15. Himi T, Iarashi M, Kataura A, Alford BR: Temporal bone findings in craniodiaphyseal dysplasia. Auris Nasus Larynx. 20: 255-261, 1993.

16. Janssens K, Thompson E, Vanhoenacker F et al: Macrocephaly and sclerosis of the tubular bones in an isolated patient: a mild case of craniodiaphyseal dysplasia? Clin Dysmorphol 12: 245-250, 2003.

17. Joseph R, Lefebvrew J, Guy E, Job JC: Dysplasie cranio-diaphysaire progressive. Ses relations avec ia dysplasie diaphysaier progessive de Camurati-Engelmann. Ann Radiol 1: 477-90, 1958.

18. Kaitila II, Stewart RE, Landow E et al: Craniodiaphyseal dysplasia. Birth Defects 11: 350-362, 1976.

19. Levy MH, Kozlowski K: Craniodiaphyseal dysplasia: report of a case. Australas Radiol 31: 431-435, 1987.

20. Little RD, Carulli JP, Del Mastro RG et al: A mutation in the lDL receptor-related protein 5 gene results in the autosomal dominant high-bone-mass trait. Am J Hum Genet 70: 11-19, 2002.

21. MacPherson RI: Craniodiaphyseal dysplasia: a disease or group of diseases? Canad Assoc Radiol 25: 22-33, 1974.

22. Makita Y, Nishimura G, Ikegawa S et al: Intrafamilial phenotypic variability in Engelmann Disease (ED): Are ED and Ribbing disease the same entity? Am J Med Genet 91: 153-156, 2000.

23. Mankin HJ: Camurati-Englemann Disease (Progressive Diaphyseal Dysplasia). In Pathophysiology of Orthopaedic Diseases. Rosemont IL, 2009, 133-137.

24. Marden FA, Wippold FJ 2nd: MR imaging features of craniodiaphyseal dysplasia. Ped Radiol 34: 167-170, 2004.

25. McHugh DA, Rose GE, Garner A: Nasolacrimal obstruction and facial bone histopathology in craniodiaphyseal dysplasia. Br J Opthalmol 78: 501-503, 1994.

26. McKeating JB, Kershaw CR: Craniodiaphyseal dysplasia. Partial suppression of osteoblastic activity in the severe progressive form with calcitonin therapy. JR Nav Med Serv 73: 81-93, 1987.

27. Naique S, Laheri VJ: Stenosis of the cervical canal in craniodiaphyseal dysplasia. J Bone Joint Surg 83B:328-331, 2001.

28. Richards A, Brain C, Dillon MJ, Bailey CM: Craniometaphyseal and craniodiaphyseal dysplasia: head and neck manifestations and management. J Laryngol Otol. 110: 328-338, 1996.

29. Schaeffer B, Stein S, Oshman D et al: Dominantly inherited craniodiaphyseal dysplasia: a new craniotubular dysplasia. Clin Genet 30: 381-391,1986.

30. Seeger LL, Hewel KC, Yao LC et al: Ribbing's disease (multiplediaphyseal sclerosis): Imaging and differential diagnosis. AJR Am J Roentgenol 167:689-694, 1996

31. Sinow JD, Gruss JS, Roberts TS et al: Intracranial and extracranial reduction osteoplasty for craniodiaphyseal dysplasia: Cleft Palate Craniofac J 33: 284-290, 1996.

32. Spranger JW, Langer LO Jr., Wiedemann HR: Craniodiaphyseal dysplasia. in Bone Dysplasias: An Atlas of Constitutional Disorders of Skeletal Development. PhIadelphia, PA, WB Saunders Company, 1974, 319-320.

33. Stransky E, Mablangan I, Tara RT: On Paget's disease with leontiasis ossea and hyperthreosis starting in an early childhood. Ann Paediatr 199: 393-408, 1962.

34. Tucker AS, Klein L, Antony GJ: Craniodiaphyseal dysplasia: evolution over a five year period. Skeletal Radiol 1: 47-53, 1976.

35. Van Wesenbeeck L, Cleiren E, Gram J et al: Six novel missense mutations in the LDL receptor-related protein 5 (LRP5) gene in different conditions with an increased bone density. Am J Hum Genet 72: 763-7.

CHAPTER 13

Pycnodysostosis: A Rare Genetic Hyperostotic Disease

Introduction:

Pycnodysostosis was named by Pierre Maroteaux and Maurice Lamy in 1962 (28) but the disease was in fact, described years before that. The difficulty lay with the relationship and similarity to osteopetrosis, a disorder first described by Albers-Schonberg in 1904 as "marble bone disease" (27). Osteopetrosis appears to be caused by interference with osteoclast action as a result of several genetic errors, which all seem to cause RANKL failure (27). Pycnodysostosis has some clinical features, which make it different from the several forms of osteopetrosis. It is an autosomal recessive disease, which affects males and females equally usually at birth and has no ethnic predilection. The disorder results in a series of manifestations, which lead to short stature, facial, clavicular and calvarial abnormalities, dense bones, deformed hands, generalized osteosclerosis and multiple fractures. In the 1990s several authors identified the cause of the disease to be a deficiency of cathepsin K, which markedly interferes with osteoclastic action and leads to fibroblastic abnormalities.

History of pycnodysostosis:

As indicated in the Introduction, pycnodysostosis was first named that by Pierre Maroteaux and Maurice Lamy in 1962 (28) but there were some reports prior to that time which identified a rare and unusual form of dense bone disease. The earliest such description was that of Montanari in 1923 (32) who reported a case of achondroplasia with abnormalities of the clavicles, skull and digits. Blasi in 1937 (2) added a case under the heading of "osteopathia". Thomsen and Guttadauro reported another patient in 1952 (48) and Giaccai and coworkers in 1954 (19) and Muckart in 1959 (35) introduced the term cleidocranial dysostosis with osteopetrosis. In 1958 and 1960, Palmer and Thomas (39,40) added a description of a case of osteopetrosis with skull abnormalities and in 1962, Andren and coworkers (1) published a paper describing a syndrome of "osteopetrosis acro-osteolytica". In 1962 Maroteaux and Lamy named the disorder pycnodysostosis, basing this term on the Greek words *pycnos* (meaning thick or dense), *dys* (defective) and *ostosis* (bone) (28).

In 1965, the same authors decided that pycnodysostosis was the entity which caused the physical abnormalities of Henri de Toulouse-Lautrec (29). Toulouse-Lautrec was a very productive French artist, who was born in 1864 and died in 1901 (13). He was described as weak and sickly as a child and was less than 4 feet in height as an adult. He had a large skull, abnormal dentition and multiple fractures of his lower extremities (13,14). As a result of the classic article defining these abnormalities, the disease also became known then and is still occasionally referred to as the Malady of Toulouse-Lautrec (29,42). Please note that there is also another entity known as Maroteaux-Lamy disease, which is Type VI mucopolysaccharide and has no relationship to pycnodysostosis.

Genetics and biochemical disorders causing pycnodystosis:

Soon after the original discovery of the syndrome several authors commented on the high likelihood that pycnodysostosis is a rare genetic disorder. According to a report by Shuler in 1963 (45) and a very comprehensive study of 33 patients from the world literature in 1966 by Stanley Elmore (7), the disease became clearly genetic in origin and most likely is transmitted as an autosomal recessive (8). Gelb and his coworkers in 1995 (16) and Polymeropoulos et al (41) in the same year linked the disease to chromosome 1q21 by homozygosity mapping. One year later, Gelb et al defined the cause as a deficiency of cathepsin K (17,18). Cathepsin K normally cleaves native Type I and II collagen and regulates osteoclast function (10,12,15,20,21,34). If cathepsin K is reduced in amount or absent, the osteoclasts become non-functional, allowing bone to become more dense and sclerotic (3,33,49). Furthermore alterations occur in both chondroitin sulfate structure and in the relation of collagen to fibroblasts so that the disorder shows changes other than those seen in the two clinical forms of osteosclerosis (26,27).

The clinical syndrome of pycnodysostosis:

Pycnodysostosis is a rare autosomal recessive genetic entity, which occurs equally in males and females and has no recognizable ethnic frequency (6,7,9,22,30,31,42,44). The number of cases that occur in any community however, are very few and the disease is often unrecognized particularly because of its similarity to other more common disorders with increased bone density (osteopetrosis (27)) or clavicular and calvarial abnormalities (cleidocranial dysplasia (4)). The children with pycnodysostosis are born with no apparent evidence of the disease except short stature, and have a normal intellect and no difficulty with respiratory, cardiac or gastrointestinal disorders (6,22,30,31,42,44).

As the children grow older they display some unusual features (6,7,9). The skull is large with fronto-occipital bulging and an open anterior fontanel even in adults (7,11,22,30,42). The face is small. The eyes are prominent and bulging and the nose elongated and sometimes described as "parrot-like" (7,30,31,38,42). The jaw is small and the teeth are remarkable in that the deciduous teeth do not disappear so that the children and adults may show two rows of teeth (36,37,38,42). Most of the children develop short limb dwarfism and rarely reach 5 feet of height (6,7,30,31,42,47). The clavicles are shorter than normal with the lateral portion sometimes absent (6,24,42). The hands are short and the fingers deformed by short terminal phalanges (6,7,9,30,31,42). Creased skin can be noted on the dorsum of the hands and fingers (6,42).

The bony structure shows genu valgum and occasionally coxa vara (6,7,30,31,42). Kyphosis, scoliosis, increased lumbar lordosis are frequently observed (6,30). Fractures of the long bones are common and heal slowly and sometimes result in non-unions and deformities (6,7,9,30,42,43,48). The chest is narrow (30,42). Only rarely are visceral complications encountered (25).

Radiographic and histologic features:

Examination of the calvarium by X-ray shows persistence of the anterior fontanel and cranial sutures (11,30,42). The cranial bones are quite large and the facial bones small (6,7,9,11,22,38). X-rays of the skeletal system show generalized osteosclerosis, with metaphyseal undermodeling of tubular bones (6,7,9). Metaphyseal bone may be enlarged in the form of Erlenmeyer flasking but rarely to the extent seen in patients with Gaucher disease (6,7,42). Usually there is evidence for fractures, with poor healing and sometimes non-union. The acromial end of the clavicle is hypoplastic (24) and there is partial aplasia of the distal phalanges of the hands and feet (6,7,42). Roentgenographic notching of the anterior aspects of the vertebrae may be present and can be associated with spinal deformity (42,50).

Histologic studies of the bone show increased density similar to that seen in osteopetrosis (12,30,42). The medullary canal is evident within the bone and despite irregularities seems to be actively involved in hematopoesis (30,31,42). The number of osteoclasts seen in the bone is normal but their degradative action in the bone in destructive spaces such as seen in Howship's lacunae are rarely observed (6,7,30,42). Occasional sites of mineralized cartilage inclusions are seen in the thickened cortices.

Treatment of patients with pycnodysostosis:

There is no treatment currently available, which could increase the amount of cathepsin K and thus activate the osteoclasts (33,34). Osteoclasts can be activated to some extent by the administration of parathyroid hormone but since the patients are often asymptomatic and have normal life styles, there may be limited value in administration of such a potentially toxic agent. Kaplan et al have proposed some form of alteration of BMP4 to treat patients with fibrodysplasia ossificans progressiva, but there does not seem to be relationship of that disorder to patients with pycnodysostosis (23). Growth hormone has been utilized for treatment of the dwarfed children with some success (5,46). Fracture treatment is an essential part of the management protocol and sometimes treatment of the coxa vara and genu varum is necessary (6,7,42). Dental problems with double rows of teeth, fractures of the jaw and osteomyelitis of the mandible may occur in these patients and may require extensive surgical and antibiotic management (36,37,38,42).

Conclusions:

Pycnodysostosis is a very rare and unusual condition, which causes multiple abnormalities in the calvarium, dental structures and long bones but does not seem to impair intelligence and indeed normal function (including artistic painting!). The children are short but that may be effectively treated by early use of growth hormone. Ideally one would like to introduce a method to activate cathepsin K and possibly do this by genetic engineering. It is possible that altering the gene at the chromosome site 1q21 would solve the problem but at this point there is no simple method for performing such a solution. Furthermore, since the patients generally do quite well in life with only short stature, jaw problems and fractures to deal with, one would hesitate to introduce a material which may significantly alter the biochemistry of collagens I and II and thus cause other major skeletal and structural problems.

References:

1. Andren L, Dymling JF Hogeman KE, Wendeberg B: Osteopetrosis acro-osteolytica: a syndrome of ostepetrosis, acro-osteolysis and open sutures of the skull. Acta Chir Scand 124: 496-507, 1962.
2. Blasi R: Osteopatia sistematica nor riportabile ai quadri gia noti. Radiol Med Rev Mensil 24: 741-754, 1937.
3. Bossard MJ, Tomaszek TA, Thompson SK et al: Proteolytic activity of human osteoclast cathepsin K: Expression, purification, activation and substrate identification. J Biol Chem 271: 12517-12524, 1996.
4. Cooper SC, Flaitz CM, Johnston DA et al: A natural history of cleidocranial dysplasia. Am J Med Genet 104:1-6, 2001.
5. Darcan S, Akisu M, Tanikl B, Kendir G: A case of pycnodysostosis with growth hormone deficiency. Clin Genet. 50: 422-425, 1996.
6. Edelson JG, Obad S, Geiger R et al: Pycnodysostosis: orthopedic aspects with a description of 14 new cases. Clin Orthop 280: 263-276, 1992.
7. Elmore SM: Pycnodysostosis: a review. J Bone Joint Surg. 49A: 153-163, 1967.
8. Elmore SM, Nance WE, McGee BJ et al: Pycnodysostosis, with a familial chromosome anomaly. Am J Med 40: 273-282, 1966.
9. Emanii-Ahari Z, Zarabi M, Javid B: Pycnodysostosis J Bone Joint Surg 51B: 307-312, 1969.
10. Everts V, Hou WS, Rialland X et al: Cathepsin K deficiency in pycnodysostosis results in accumulation of non-digested phagocytosed collagen in fibroblasts. Calc Tissue Int 73: 380-386, 2003.
11. Figueiredo J, Reis A, Vaz R et al: Porencephalic cyst in pycnodysostosis. J Med Genet 26: 782-784, 1989.
12. Fratzl-Zelman N, Valenta A, Roschger P et al: Decreased bone turnover and deterioration of bone structure in two cases of pycnodysostosis J Clin Endocrinol Metab 89:1538-1547, 2004.
13. Frey J: Toulouse-Lautrec: a life. New York, Viking Press. 1994.
14. Frey J: What dwarfed Toulouse-Lautrec? Nature Genet 10: 128-130, 1995.
15. Fujita Y, Nakata K, Natsuo Y et al: Novel mutations of the cathepsin K gene in patients with pycnodysostosis and their characterization. J Clin Endocrinol Metab 85: 425-431, 2000.

16. Gelb BD, Edelson JG, Desnick RJ: Linkage of pycnodysostosis to chromosome 1q21 by homozygosity mapping. Nature Genet 10: 235-237, 1995.

17. Gelb BD, Shi GP, Chapman HA, Desnick RJ: Pycnodysostosis, a lysosomal disease caused by cathepsin K deficiency. Science 273: 1236-1238, 1996.

18. Gelb BD, Willner JP, Dunn TM et al: Paternal uniparental disomy for chromosome 1 revealed by molecular analysis of a patient with pycnodysostosis. Am J Hum Genet 62: 848-854, 1998.

19. Giaccai L, Salaam M, Zellweger H: Cleidocranial dysostosis with osteopetrosis. Acta Radiol 41: 417-424, 1954.

20. Hou W-S, Bromme D, Zhao Y et al: Characterization of novel cathepsin K mutations in a pro and mature polypeptide regions causing pycnodysostosis. J Clin Invest 103: 731-738, 1999.

21. Kaftenah W, Bromme D, Buttle DJ et al: Human cathepsin K cleaves native type I and II collagens at N-terminal end of the triple helix. Biochem J 331: 727-732, 1998.

22. Kajii T, Homma T, Ohsawa T: Pycndysostosis. J Pediatr 69: 131-133, 1966.

23. Kaplan FS, Fiori J, De La Pena, LS et al: Dysregulation of the BMP-4 signaling pathway in fibrodysplasia ossificans progressiva. Ann Ny Acad Sci. 1068: 54-65, 2006.

24. Karakurt L, Yilmaz E, Belhan O, Serin E: Pycnodysostosis associated with bilateral congenital psuedoarthrosis of the clavicle. Arch Orthop Traum Surg 123: 125-127, 2003.

25. Kozlowski K, Yu JS: Pycnodysostosis: a variant form with visceral manifestations. Arch Dis Child 47: 804-807, 1972.

26. Li Z, Hou WS, Escalente-Torres CR et al: Collagenase activity of cathepsin K depends on complex formation with chondroitin sulfate. J Biol Chem 277: 28669-26676, 2002.

27. Mankin HJ: Osteopetrosis. In Pathophysiology of Orthopaedic Diseases. Rosemont, IL, American Academy of Orthopaedic Surgeons, 2006, 123-130.

28. Maroteaux P, Lamy M: La pycnodysotose. Presse Med 70: 999-2002, 1962.

29. Maroteaux P, Lamy M: The malady of Toulouse-Lautrec. JAMA: 1919-715-117, 1965.

30. Meredith SC, Simon MA, Laros GS, Jackson MA: Pycnodysostosis: a clinical, pathological and ultramicroscopic study of a case. J Bone Joint Surg 60A: 1122-1128, 1978.

31. Mills KLG, Johnston AW: Pycnodysostosis. J Med Genet 25: 550-553, 1988.

32. Montanari UI: Achondroplasia e disostos cleido-cranica-digitale. Chi Orgai Movimento 7: 379-391, 1923.

33. Motyckova G, Fisher DE: Pycnodysostosis: role and regulation of cathepsin K in osteoclast function and human disease. Curr Mol Med 2: 407-421, 2002.

34. Motyckova G, Weilbaecher KN, Horstmann M, et al: Linking osteopetrosis and pycnodysostosis: regulation of cathepsin K expression by the micropthalmia transcription factor family. Proc Nat Acad Sci US 98: 5798-5803, 2001.

35. Muckart RD: Cranio-cleido dysostosis. J Bone Joint Surg 41B 633, 1959.

36. Muto T, Yamazaki A, Takeda S et al: Pharyngeal narrowing as a common feature in pycnodysostosis-a cephalometric study. Int J Oral Maxillofac Surg. 34: 680-685, 2005.

37. Norholt SE, Bjerregaard J, Mosekilde L: Maxillary distraction osteogenesis in a patient with pycnodysostosis: a case report. J Oral Maxillofac Surg 62: 1037-1040, 2004.

38. O'Connell AC, Brennan MT, Francomano CA: Pycnodysostosis: orofacial manifestations in two pediatric patients. Pediatric Dentistry. 20: 204-207, 1998.

39. Palmer PES: Case Report: Osteopetrosis with multiple epiphyseal dysplasia. British J Radiol 31: 455-457, 1960.

40. Palmer PES, Thomas JEP Case reports: Osteopetrosis with unusual changes in the skull and digits. British J Radiol 31: 705-708, 1958.

41. Polymeropoulos MH, Ortiz de Luna RI, Ide SE et al: The gene for pycnodysostosis maps to human chromosome 1cen-q21. Nature Genet 10: 238-239, 1995.

42. Rimoin DL, Lachman RS: Genetic disorders of the osseous skeleton. In Mckusick's Heritable Disorders of Connective Tissue. Fifth Edition, Beighton P, editor. St. Louis, Mosby, 1993 656-657.

43. Roth VG: Pyknodysostosis presenting with bilateral subtrochanteric fractures: case report. Clin Orthop 117: 247-253, 1976.

44. Sedano HP, Gorlin RJ, Anderson VE: Pycnodysostosis: clinical and genetic considerations. Am J Dis Child 116: 70-77, 1968.

45. Shuler SE: Pycnodysostosis. Arch Dis Child. 38:620-625, 1963.

46. Soliman AT, Rajab A, Al Salmi I et al: Defective growth hormone secretion in children with pycnodysostosis and improved linear growth after growth hormone treatment. Arch Dis Child 75: 242-244, 1996.

47. Taylor MM, Moore TM, Harvey JP Jr.: Pycnodysotosis: a case report. J Bone Joint Surg 60A: 1128-1130, 1978.

48. Thomsen G, Guttadauro M: Cleidocranial dysostosis associated with osteosclerosis and bone fragility. Acta Radiol 37:559-567,1952.

49. Troen BR: The role of cathepsin K in normal bone resorption. Drug News Perspect 17: 19-28, 2004.

50. Zenke MS, Hatori M, Tago S et al: Pycnodysostosis associated with spondylolysis. Arch Orthop Trauma Surg 122: 248-250, 2002.

CHAPTER 14

Idiopathic Hyperphosphatasia, Bakwin Eiger Syndrome

Idiopathic hyperphosphatasia, otherwise known as juvenile Paget's disease is a rare disorder first clearly identified by Bakwin and Eiger in 1956 (1). It is a genetic disorder affecting both sexes and producing in early life, a characteristic pattern of cortical bone hyperplasia, enlargement of the skull, thickening and bowing of the long bones and an array of clinical findings which cause severe impairment and poor survival for affected children (2,19). The cause of the disorder is believe to be a mutation in the gene TNFRS11B encoding osteprotegerin (OPG) which leads to alteration RANK, RANK-L role in the production of osteoclasts and this causes a marked increase in the metabolism of the bone (6, 20). The descriptive term hyperphosphatasia is related to the marked increase in the bone specific alkaline phosphatase, also seen in elderly persons with Paget's disease (5).

Nomenclature and history: The disease is very uncommon and there has been a moderate confusion in relation the nomenclature. Harry Bakwin and Marvin Eiger published and article entitled "Fragile bones and macrocranium" in the Journal of Pediatrics in 1956 in which they described a child who had fusiform swelling and bowing of the long bones and enlargement of the calvarium (1). There were other aspects including the elevation of the serum alkaline phosphatase and both the osseous and biochemical findings resembled the changes seen Paget's disease (5,9). The latter disorder is considerably more frequent, occurs much later in life and is of uncertain etiology, with both genetic errors and virus infection having suspected but never proven. The result is that the name "Juvenile Paget's Disease" was introduced for this entity, along with names chronic hyperphosphatasia, chronic osteopathy-hyperphosphatasia, hyperostostosis corticalis deformans juvenilis, hyperphosphatasia-osteoectasia syndrome and familial osteoectasia (8,9,12,18).

Following the report by Bakwin and Eiger in 1956 describing a family of Puerto Rican ancestry with children with the disease, Swoboda in 1958 reported two sisters with the disease and described it as hyperostosis corticalis deformans juveniles (2). Choremis et al in the same year reported additional cases (5) and Caffey in 1961 (3). Fanconi et al in 1964 reported the X-ray and histologic changes in a Brazilian male child and suggested the name osteochalasia desmalis familiaris (8). Eyring and Eisenberg in 1968 reported a brother and a sister (7) and reports by Golob et al in 1996 defined the sometimes extraordinary increase in the bone specific serum alkaline phosphatase (9). In 1996, Whyte et al identified the genetic error in TNFRS11B gene (17) and Cundy in the same year was able to establish the relationship of the disease to the error in the synthesis of osteoprotegerin (15).

Biologic causation of idiopathic hyperphosphatasia: In terms of genesis, the disease appears to be caused by a truncating TNFRSF11B mutation on chromosome 8q24.2, which causes a deficiency of osteoprotegerin (OPG) (16). OPG is closely tied to the system which result in synthesis of osteoclasts, specifically RANK (receptor activator of nuclear factor kB) and RANK-ligand (RANKL). Under normal circumstances, OPG acts to suppress bone turnover by functioning as a decoy receptor for RANKL (20). The absence of OPG allows RANKL to activate RANK to markedly increase the synthesis of osteoclasts by activation of hematopoietic precursor cells. This increases the number of osteoclasts, which then results in alterations of the bones based on the coupling of osteoblastic synthesis and osteoclastic resorbtion and also allows the production of excessive amounts of bone-specific alkaline phosphatase (4,19,20).

The clinical syndrome for idiopathic hyperphosphatasia: There is little doubt based on a number of reports that juvenile Paget's disease is genetic and appears to be autosomal recessive in transmission. The syndrome however has an array of presentations in very young children with very severe bone and skull disorders but may be virtually asymptomatic in some adults with minimal bone changes and only minor increases in serum alkaline phosphatase. The disease appears to be slightly more frequent in females than in males and cases from South and Central America and India have been reported (3). The studies by Cundy et al (6)describing patients from New Zealand and Australia suggest that there is increased frequency in those countries which is interesting since the studies by Barry some years ago suggest that Australia is a country with the highest frequency of standard Paget's disease and Paget's sarcoma.

Clinically children with familial hyperphosphatasia are generally of short stature and display muscular weakness (13). They only rarely have mental retardation. The head is larger than normal and imaging studies show marked thickening of the calvarium with islands of increased density (2). Hearing loss becomes worse with advancing years. Dental abnormalities are common with loose and abnormally shaped teeth and reports have suggested that the children may have blue sclerae and also retinal degeneration related to angiod streaks (4,18,19). The bones generally show a modest osteopenia and expanded structure and frequent bowing deformities (10,19). Pelvic deformities may be present along with vertebral abnormalities and sometime collapsed segments. Fractures are common and the bones are described as fragile. The fractures and the bowing lead to increased skeletal deformity and some of the patients are markedly disabled as they get older. Imaging studies are similar to those seen in some patients with early active Paget's disease and the bone scan is virtually always positive (4,8,13,18).

Laboratory data show increased serum alkaline phosphatase and especially bone specific alkaline phosphatase (7). Serum acid phosphatase is also elevated and the patients frequently show elevated phosphate levels, increased serum uric acid, increased hydroxyproline and periodic episodes of hypercalcemia (7).

Histologic studies show similar changes to active Paget's disease with marked increase in the number of osteoclasts destroying bone coupled with increased osteoblastic activity, but the bone is disordered in structure and alignment (7). Stress fractures are common and woven bone may be present rather than mature cortical structural bone.

Treatment of idiopathic hyperphosphatasia:

Until relatively recently there was no sensible treatment protocol for patients with this syndrome. Fractures required fixation and repairs often operative and the children were disabled and psychiatrically damaged. Visual and hearing disturbances made education difficult. A great contribution was the introduction of calcitonin which markedly reduced the rate of bone destruction and for many patients resulted in fewer fractures and less deformities and disability (14). The use of inhalation administration also made life simpler for young children. A second treatment protocol appeared when the bisphosphonates were introduced and these agents markedly reduced bone destruction. Cyclic intravenous pamidronate was introduced and over several months decreased bone deformity and the level of alkaline phosphatase (11) Most recently recombinant osteoprotegerin administered subcutaneously once weekly has been utilized and appears to be effective in increasing bone structural mass and reducing the biochemical findings in he serum.

Hyperphosphatasia references:

1. Bakwin H, Eiger MS: (1956). Fragile bones and macrocranium. J Pediatr 49:558-564.
2. Bakwin H, Golden A, Fox S: (1964). Familial osteoectasia with macrocranium. Am J Roentgen 91: 609-617.
3. Caffey JP: Pediatric X-ray Diagnosis (1961). Chicago, Year Book Med Publ.
4. Caffey JP: (1972). Familial hyperphosphataemia with ateliosis and hypermetabolism of growing membranous bone: review of the clinical radiographic and chemical features Bull Hosp Joint Dis. 81-110.
5. Choremis C, Yannakos D, Papadatos C, Baroutsou E: (1958) Osteitis deformans (Paget's disease) in an 11 year old boy. Helv Paediat Acta 13: 185-188.
6. Cundy T, Hegde M, Naot D et al: (2002). A mutationin the gene TNFRSF11B encoding osetoprotegerin causes an idiopathic hyperphosphatasia phenotype. Hum Molec Genet. 11: 2119-2127.
7. Eyring EJ, Eisenberg E: (1968). Congenital hyperphosphatasia. A clinical pathological and biochemical study of two cases. J Bone Joint Surg 50A: 1099-1117.
8. Fanconi G, Oroeira G, Uehlinger E Giedion (1964) Osteochalasia desmalis familiaris. Hyperphosphatasia and macrocranium Helv Pediatr Acta 19: 279-295.
9. Golub DS, McAlister WH, Mills BG et al: (1996) Juvenile Paget disease: lifelong features of a mildly affected young woman. J Bone Miner Res. 11: 132-142.
10. Iancu TC, Almagor G, Friedman E et al: (1978). Chronic familial hyperphosphatemia. Radiology 129: 669-676.
11. Spindler A, Berman A, Maualen C et al: (1992), Chronic idiopathic hyperphosphatasia: report of a case treated with pamidronate and a review of the literature. J Rheum 19: 642-645.
12. Stemmermann GN: (1966). An histology and histochemical study of familial osteoectasia (chronic idiopathic hyperphosphatasia), Am J Pathol 48: 641-651.

13. Thompson RC Jr., Gaull GE, Horwitz J, Schenk RK: (1969). Hereditary hyperphosphatasia. Study of three siblings. Am J Med 47: 209-219.

14. Whalen JP, Horwith M, Krook L et al: (1977). Calcitonin treatment of in hereditary bond dysplasia with hyperphosphatasemia: a radiographic and histologic study of bone. Am J Roentgen. 129: 29-35.

15. Cundy T, Davidson J, Rutland MD et al: (2005) Recombinant ostoprotegerin for juvenile Paget's disease. N Engl J Med 353: 918-923.

16. Hofbauer LC, Schoppet MP, Whyte MN et al: (2002) Osteoprotegerin deficiency and juvenile Paget's disease. N Engl J Med 347: 1622-1623.

17. Whyte MP, Hughes AE (2002). Expansile skeletal hyperphosphasia is caused by a 15 base pair tandem duplication in TNFRSF11A encoding RANK and is allelic to familial expansile osteolysis. J Bone Miner Res 17: 26-29.

18. Whyte MP, Mills BG, Reinus WR: (2000). Expansile skeletal hyperphosphatasia: a new familial metabolic bone disease. J Bone Miner Res 15: 2330-2344.

19. Cassinelli HR, Mautalen CA, Heinrich JJ et al: (1992) Familial idiopathic hyperphosphatasia (FIH) Bone Miner 19:175-184.

20. Janssens K, deVernejoul MC, de Freitas F (2005) An intermediate form of juvenile Paget's disease caused by a truncating TNFRSF11B mutation. Bone 36: 542-548.

CHAPTER 15

Multiple Epiphyseal Dysplasias

Multiple epiphyseal dysplasias (MEDs) are a group of genetic disorders characterized by alterations of epiphyseal cartilage growth and function. These changes are genetic, are apparently related in part to alterations in the form of collagen type IX, which cause a series of structural alterations in the bones, mostly of the extremities and less commonly the spine. The patients may be of only moderately short stature and have normal mentation but have sometimes disabling and deforming alterations in bones and especially the joints. Most often the disorders are transmitted as autosomal dominant although there is also a recessive form. Early sometimes severe osteoarthritis of the hip and knee are characteristic along with short digits in the hands and feet. A mild form of the disorder was described by Ribbing in 1937, and Fairbank described a more severe form in 1947. In addition there are a number of similar entities with variable alterations and sometimes disabling characteristics.

Nomenclature and history:

Part of the difficulty with definition of MED is its close approximation both clinically and genetically with some other disorders, namely pseudochondrodysplasia ((PSACH) and Wallcot-Rallison syndrome (MED with early onset diabetes). In addition however there are the other names for the entity including Fairbank's disease, Ribbing's disease, Muller-Ribbing's syndrome, dysplasia epiphysialis multiplex, dysostosis enchondral is polyepiphysaria, polyosteochondrosis, epiphyseal dysostosis, hereditary enchondral dysostosis and chondrodystrophia calcificans congenita.

It was Barrington-Ward who in 1917 (4) first described bilateral coxa-vara deformities in a brother and a sister and expressed the opinion that the disorder resulted from a genetic error. In 1937, Seved Ribbing, a Swiss radiologist (51) described a mild form of epiphyseal alteration in structure and function characterized by shorter bones and structural change. Two years later, Walther Muller of Germany (45) provided a similar description, so that the milder form became known as Muller-Ribbing syndrome. In 1946, Sir Thomas Fairbank of England (16) described a more aggressive form of the epiphyseal abnormality, which resulted in major issues in terms of epiphyseal contribution to bone growth and the integrity of the joint structures. In 1954, Jackson and

Hanelin, working with Fuller Albright (32) pointed out that a number of disorders characterized by epiphyseal abnormalities exist and then described some of them in some detail. Reports by Waugh in 1952(63), by Maudsley in 1955 (41), and by Odman in 1959 (48) described families with the disorder and established the dominant characteristic of the genetic transmission. In 1958, Freiberger (17) reported the radiologic changes in three patients. Elsbach in 1959 (14) further defined the nature of the hip disease in patients with the disorder. In 1962, Hodkinson (24) first described the characteristic "double patella" which is virtually diagnostic for MED and Hulvey and Keats in 1969 (26) defined the spinal changes seen in some of these patients. Over the next decade, further reports by Hunt and colleagues in 1967 (27), Koslowski and Lipsky in the same year (35), Juberg and Holt in 1968 (33), Monsoor in 1970)(40) and Gamboa and Lisker in 1974 (18) detailed the clinical characteristics, the radiographic changes, and the histologic abnormalities for the patients with both mild and severe multiple epiphyseal dysplasia. In 1972, Wolcott and Rallison first described a patient with infancy-onset diabetes mellitus associated with multiple epiphyseal dysplasia (62). In 1973, Joseph Bailey (1) published a remarkable text-book entitled "Disproportionate Short Stature" in which he describes in great detail, patients with multiple epiphyseal dysplasia along with a number of other similar disorders.

Pathophysiology and genetics for multiple epiphyseal dysplasia:

The pathologic changes which occur in patients with multiple epiphyseal dysplasia are initially confined to the epiphyseal plates, mostly of the long bones (1,27,53). Instead of normal epiphyseal structure with clearly defined zonal cellular organization, the epiphyses become deranged and distorted with resultant disturbance of the linear organization of the cells, the presence of a lamina dura and a structured calcified zone (1,27,53). The epiphyses become enlarged, extend further into the underlying bone and lose some of their capacity to increase the bone length (37,53). Even with the milder Ribbing form of the disease, somewhat later in the course, similar changes occur in the articular cartilage and the patients may develop a form of progressive osteoarthritis at a sometimes early age (10,53). Histologic features show considerable distortion of the structure of the epiphyses and subsequently for many patients, the articular cartilage as well (1,27). The bones that develop as a result of epiphyseal chondrification are at first normal but relatively early for Fairbank's severe form of the disorder, become not only short but may become distorted in their structure (1,10,27,53) Shortening and bowing of limbs, coxa vara deformities and a disorder resembling Perthes disease are commonly seen as the children grow older (1,10,17,23,27,52,53).

The cause of these disorders is believed to be a biochemical genetically induced failure to develop normal collagen in the cartilage and most of the changes are believed to be due to alterations in type IX collagen (6,8,10,25,36,43,46). Type IX collagen is a nonfibrial structure consisting of 3 collagenous domains (COLI-COL3) interrupted by 4 noncollagenous domains (NCI-NC4) (6,10,25,60). Type IX collagen is a structural component for hyaline cartilage, intervertebral disks and the vitreous body of the eye (10,25,46,53). The genetic errors, which cause structural changes in Type IX collagen appear to lie in multiple chromosomal sites and involve five separate mutations. The first and most frequently encountered error in patients with multiple epiphyseal dysplasia as well as in those with pseudochondrodysplasia occurs in cartilage oligiomeric matrix protein (COMP) (8,9,21,22,25,28,34,36,58). The chromosomal locus for this error is i 9p 13.1 and the errors consist of both missense and deletions. Some of the additional mutations which are believed to be the cause of principally multiple epiphyseal dysplasia occur in COL9Al (EDMl) at chromosomal locus 6q13 (II); COL9A2 (bDM2) at locus Ip33-p32.2 (46); COL9A3 (bDM3) at locus 20q13.3 (47,49); diastrophic dysplasia sulfate transporter (bDM4) at locus 5q32-q33(2,38); and matrilin-J (bDM5) at locus 2p24-p23 (7,31,37,39,44). Multiple errors have been reported and the estimate is as high as 50 occurring in these sites, ail of which lead to skipping of an exon in the Type IX collagen COL3 domain (13,15,30). The COL3 domain is functionally important and appears to mediate interaction between Type IX collagen and other components of the cartilage extracellular matrix and more specifically Type II collagen. Although most of these errors occur in patients with the autosomal dominant form of the disorder, an error in the diastrophic dysplasia sulfate transporter gene (DDTST) seems to be the cause of autosomal recessive disease (2,38).

The clinical syndrome of multiple epiphyseal dysplasia:

Multiple epiphyseal dysplasia comes in two forms: the mild as described by Ribbing (51,52,53); and the more aggressive or "severe" form as described by Fairbank (16,53).

Both forms present early in childhood, but the mild form is less evident throughout the first 2 or more years of life (10,53). Most often the diseases are transmitted as autosomal dominant disorders and occur with equal frequency in males and females (1,10,53).

There does not seem to be any specific ethnic frequency for either form of the disorder (1,53).

Children most frequently appear normal at birth and have no evidence of mental retardation, musculoskeletal or neural alterations in early life. The most evident findings seen early in Fairbank's disorder and somewhat later in Ribbings disease are bowing and some shortening of the extremities (10,12,20). This appears to more frequently affect the femur and especially the knees and hips which may be limited in motion and somewhat deformed in appearance as the child ages (20,37,53,55). Coxa vara is a common deformity and the hands and feet are "stubby" and the digits shorter than normal, especially for the severe form of the disease (1,20,53,55).

When children begin to walk they may have a waddling gait and may present with a decreased range of motion in the spine (10,12,20,26,40). Bowing abnormalities may occur in the forearm and especially the humerus and the shoulders may show a decreased range of motion (12,29). As the children age, particularly those with severe disease, they often show a decreased height as compared with other children and although they do not appear to be "dwarfs", they for the most part are under 60 or 65 inches in height (1,53,55).

Later in life, the children show some marked deformities, especially if they have the severe form of the disease. Perthes disease may be present in the hips and in some cases dislocations of the hips can occur (23,55). Another most unusual feature is a patellar abnormality, which on imaging is defined as "double layered" (24,50,54,57). The patellas are small and may become chronically dislocated. Spinal problems often occur with advancing age and show narrowing of the disk spaces on imaging and decreased vertebral body height (26,55) which sometimes cause symptoms. Some patients may show macrocephaly and frontal bossing and a small number may show some neuromuscular anomalies (5).

As the patients age they very frequently develop early osteoarthritis of the knees, ankles, hips and elbows along with shortening of the digits and limitation of motion in the carpus and the ankle (29,40,42,55,61). Although these changes may affect their functional lives most of the milder patients are still able to carry on most normal activities with only moderate disability. The severe forms of the disease patients are more disabled and often require corrective surgery.

Imaging studies have shown some remarkable changes. In a comprehensive study by Miura et al (42), some of the more common changes on X-ray studies of the knee included a shallow femoral trochlear groove, flat intracondylar eminence, depression of the lateral tibial plateau, genu valgum and early osteoarthritic change. The cardinal feature of the disease consists of a double-layered patella (19,24,50,54,57,61). The patella is small and has two layers and is often laterally displaced. The height of the vertebral segments is decreased and the disk space somewhat narrowed (26,53). As the patient becomes older some spinal osteoarthritic changes can occur. The hips often show coxa vara in young children and may become displaced as the child ages (1,10,14,17,59).

Osteoarthitic and Perthes-like alterations occur early in the patient's life and may be very disabling (23,55,59). Hands and feet are short and the carpal and tarsal bones are often distorted in shape and structure. The joint spaces for the metacarpals and phalanges are markedly thinned in the severe form of the disease (1,53,55,59).

It is important to note that patients with mild disease may have only minor or limited changes on imaging and in terms of clinical function, while those with severe disease may be quite short in stature and have multiple deformities of the spine and extremities and are functionally limited and disabled (3,12,20,23,26,29,42,53,55,59,61).

Treatment of patients with multiple epiphyseal dysplasia:

It is sometimes difficult to establish the diagnosis of MED particularly the mild form and for many of these patients medical treatment may not be necessary. Review of family history is important to help establish the diagnosis and imaging studies of the knees, hips, spine, shoulders, hands and feet at 2 or more years of age may be helpful in establishing the diagnosis. Studies of the patella on X-ray, CT or MRI may help to establish the diagnosis, since the double patella image is virtually diagnostic for this disorder. In view of the occasional occurrence of Wolcott-Rallison syndrome, it is appropriate to test even young children for sometimes severe diabetes (56,62). Genetic analyses are helpful in distinguishing multiple epiphyseal dysplasia from pseudoachondroplasia and other disorders in which limb shortening, bowing and joint abnormalities are present (8,9,10,60).

Currently there are no treatments designed to correct the genetic errors and no approach is being utilized to improve the function and status of collagen type IX. Treatment is designed to effectively correct any of the anatomical abnormalities that may occur particularly with the severe form ofthe disorder (3,20,26,29,42). Thus physiotherapy and exercise approaches may be helpful and obesity should be avoided (12). Psychologic and family support systems are important particularly for patients with severe disease.

Bowing deformities may be corrected by osteotomy and hip, knee and shoulder problems can be treated by surgical approaches including at times the utilization of metallic implants

(19,29,42,55,61). Osteotomies of the hip are sometimes helpful in correcting coxa vara and in treating the Perthes-like disorders (53,55). Most patients live a long life, even with severe disease and have no real problems with neurologic, cardiac or respiratory function.

Conclusions:

Multiple epiphyseal dysplasia represents a complex genetic disorder for which the gene errors have been fairly well established but the clinical entities are currently difficult to evaluate and especially to treat. There is little doubt that the disorder is caused by genetic errors of several sources most of which affect collagen type IX, which is involved in cartilage structure and metabolism. There are many errors described but basically they lead to abnormal epiphyseal cartilage structure in the newborn child, which over time causes a variety of orthopaedic problems. The bones and joints are affected which leads to diminished height, joint abnormalities, growth disturbances in the long bones and the spine. The children are mentally intact and except for some unusual forms of the disease, have no real neurologic or metabolic problems.

The problem with the disorder is that there are a number of forms with different presentations. The severe form causes bowing, sometimes severe hip, shoulder and knee problems and spine and hand and foot deformities. The mild form has similar effects but by comparison, they are often minimal and not really requiring treatment. Now that we understand the genetic errors involved, it would be of great value to attempt to develop a gene alteration protocol, which would prevent the severe form from causing disability and improve the lives of the mild form. Can that be done? It seems reasonable enough and the risks are probably minimal, so it should be attempted.

References:

1. Bailey JA II: Disproportionate Short Stature: Diagnosis and Management. Philadelphia, WB Saunders Company, 1973,380-437.
2. Ballhausen D, Bonafe L, Terhal P et al: Recessive multiple epiphyseal dysplasia (rMED): phenotype delineation in eighteen homzygotes for DTDST mutation R279W. J Med Oenet 40: 65-71,2003.
3. Bajuifer S, Letts M: Multiple epiphyseal dysplasia in children: beware of overtreatment! Can J Surg 48: 106-109,2005.
4. Barrington-Ward LE: Double coxa vara with other deformities occurring in brother and sister. Lancet I: 157-159, 1912
5. Bondestam J, Pihko H, Vanhanen SL et a1: Skeletal dysplasia presenting as a neuromuscular disorder-report of three children. Neuromuscul Disord 17: 231-234, 2007.
6. Bonnemann CO, Cox OF, Shapiro F et al: A mutation in the alpha 3 chain of type IX collagen causes autosomal dominant multiple epiphyseal dysplasia with mild myopathy. PNAS 97: 1212-1217,2000.
7. Borochowitz ZU, Scheffer D, Adir BV et al: Spondylo-epi-metaphyseal dysplasia (SEMD) matrilin 3 type: homozygote matrilin 3 mutation in a novel form of SEMD. J Med Genet 41: 366-372, 2004.

8. Briggs MD, Chapman KL: Pseudoachondroplasia and multiple epiphyseal dysplasia: mutation review, molecular interactions and genotype to phenotype correlations. Hum Mutat 19: 465-478, 2002.

9. Briggs MD, Hoffman SM, King LM et al: Pseudoachondroplasia and multiple epiphyseal dysplasia due to mutations in the cartilage oligomeric matrix protein gene. Nat Genet 10: 330-336, 1995.

10. Chapman KL, Briggs MD, Mortier GR: Review: clinical variability and genetic heterogeneity in multiple epiphyseal dysplasia. Pediatr Pathol Mol Med 22: 53-75,2003.

11. Czarny-Ratajczak M, Lohiniva J, Annunen S et al: A mutation in COL9Al causes multiple epiphyseal dysplasia: further evidence for locus heterogenieity. Am J Hum Genet 69: 969-980, 2001.

12. Damignani R, Young NL, Cole WO et al: Impairment and activity limitation associated with epiphyseal dysplasia in children. Arch Phys Med Rehabil85: 1647-1652, 2004.

13. Deere M, Saford T, Francomano CA et al: Identification of nine novel mutations in cartilage oligomeric matrix protein in patients with psuedoachondroplasia and multiple epiphyseal dysplasia. Am J Med Genet 85: 486-490, 1999.

14. Elsbach L: Bilateral hereditary micro-epiphyseal dysplasia of the hips. J Bone Joint Surg 41B: 514-523, 1959.

15. Eyre S, Roby P, Wolstencroft K et al: Identification of a locus for a form of spondylepiphyseal dysplasia on chromosome 15q26.1: exclusion of aggrecan as a candidate gene. J Med Genet 39: 634-638, 2002.

16. Fairbank HAT: Dysplasia epiphysialis multiplex. Br J Surg 135: 225-232, 1947.

17. Freiberger RH: Multiple epiphyseal dysplasia: report of three cases. Radiology 70: 379-385, 1958.

18. Gamboa I, Lisker R: Multiple epiphyseal dysplasia tarda, A family with autosomal recessive inheritance. Clin Genet 4: 15-19, 1974.

19. Gardner J, Woods D, Williamson D: Management of double-layered patellae by compression screw fixation. J Pediatr Orthop 8: 39-41, 1999.

20. Haga N, Nakamura K, Takikawa K et al: Stature and severity in multiple epiphyseal dysplasia. J Pediatr Orthop 18: 394-397, 1998.

21. Hecht JT, Hayes E, Haynes R Cole WG: COMP mutations, chondrocyte function and cartilage matrix. Matrix Bioi 23: 525-533,2005.

22. Hedbom E, Antonsson P, Hjerpe A et al: Cartilage matrix proteins. An acidic oligomeric protein (COMP) detected only in cartilage. J Bioi Chern 267: 6132-6136, 1992.

23. Hesse B, Kohler G: Does it always have to be Perthes' disease? What is epiphyseal dysplasia. Clin Orthop 414: 219-227,2003.

24. Hodkinson JM: Double patellae in multiple epiphysial dysplasia. J Bone Joint Surg 44B: 569-572, 1962.

25. Holden P, Meadows RS, Chapman KL et al: Cartilage oligomeric matrix protein interacts with type IX collagen, and disruptions to these interactions identify a pathogenetic mechanism in a bone dysplasia family. J Bioi Chem 176: 6046-6055, 2001.

26. Hulvey JT, Keats T: Multiple epiphyseal dysplasia: a contribution to the problem of spinal involvement. Am 1. Roentgenoll 06: 170-177, 1969.

27. Hunt DD, Ponseti IV, Pedrini-Mille A, Pedrini V: Multiple epiphyseal dysplasia in two siblings: histological and biochemical analyses of epiphyseal plate cartilage in one. J Bone Joint Surg 49A: 1611-1627, 1967.

28. Ikegawa S, Ohashi H, Nishimura G et al: Novel and recurrent COMP (cartilage oligomeric matrix protein) mutations in pseuoachondroplasia and multiple epiphyseal dysplasia. Hwn Genet. 103: 633-638, 1998.

29. Ingram RR: The shoulder in multiple epiphyseal dysplasia. J Bone Joint Surg 73B: 277-279, 1991.

30. Itoh T, Shirahama S, Nakashima E et al: Comprehensive screening of multiple epiphyseal dysplasia mutations in Japanese population. Am J Med Genet. 140: 1280- 1284,2006.

31. Jackson GC, Barker FS, Jakkula E et al: Missense mutations in the beta strands of the single A-domain of matrilin-3 result in multiple epiphyseal dysplasia. J Med Genet 41: 52-59, 2004.

32. Jackson WPU, Hanelin J, Albright F: Metaphyseal dysplasia, epiphyseal dysplasia, diaphyseal dysplasia and related conditions. II Multiple epiphyseal dysplasia: its relation to other disorders of epiphyseal development. AMA Arch Int Med 94: 886-901, 1954.

33. Juberg RC, Holt JF: Inheritance of multiple epiphyseal dysplasia tarda. Am J Hum Genet. 20: 549-563, 1968.

34. Kennedy J, Jackson G, Ramsen S et al: COMP mutation screening as an aid for the clinical diagnosis and counseling of patients with a suspected pseudoachondroplasia or multiple epiphyseal dysplasia. Eur J Hum Genet 13: 547-555,2005.

35. Koslowski K, Lipska E: Hereditary dysplasia epiphysealis multiplex. Clin Radiol 18: 330-336, 1967.

36. Lachman RS, Krakow D, Cohn DH, Rimoin DL: MED, COMP, multilayered and NEIN: an overview of multiple epiphyseal dysplasia. Pediatr Radiol 35: 116-123, 2005.

37. Makitie 0, Mortier GR, Czarny-Ratajczak Met al: Clinical and radiographic finds in multiple epiphyseal dysplasia caused by MATN3 mutations: description of 12 patients. Am J Med Genet A 125: 278-284,2004.

38. Makitie 0, Savarirayan R, Bonafe K et al: Autosomal recessive multiple epiphyseal dysplasia with homozygosity for C653S in the DTDST gene: double-layer patella as a reliable sign. Am J Hum Genet 65: 31-38, 1999.

39. Mann 11, Ozbek S, Engel J et al: Interactions between the cartilage oligomeric matrix protein and matrilins, Implications for matrix assembly and the pathogenesis of chondrodysplasias. J BioI Chem 279: 25294-25298, 2004.

40. Mansoor IA: Dysplasia epiphysealis multiplex. Clin Orthop 72: 287-292, 1970.

41. Maudsley RH: Dysplasia epiphysealis multiplex: a report of fourteen cases in three families. 1 Bone Joint Surg 37B: 228-240, 1955.

42. Miura H, Noguchi Y, Mitsuyasi H et al: Clinical features of multiple epiphyseal dysplasia expressed in the knee. Clin Ortho 380: 184-190, 2000.

43. Mortier GR, Chapman K, Leroy JL, Briggs MD: Clinical and radiographic features of multiple epiphyseal dysplasia not linked to the COMP or type IX collagen genes. Eur 1 Hum Genet. 9: 606-612,2001.

44. Mostert AJ, Dijkstra PF, Jansen BR et al: Familial multiple epiphyseal dysplasia due to a matrilin—3 mutation: further delineation of the phenotype including 40 years follow- up. Am J Med Genet 120A: 490-497, 2003.

45. MUller W: Das bild der multiplen erblichen storung der epiphsenverknocherung. Zachr Orthop 69: 257, 1939.

46. Muragaki Y, Mariman EC, van Beersum SE et al: A mutation in the gene encoding the alpha 2 chain of the fibril-associated collagen IX, COL9A2, causes multiple epiphyseal dysplasia (EDM2), Nat Genet 12: 103-105, 1996.

47. Nakashima E, Kitoh H, Maeda K et al: Novel COL9A3 mutation in a family with multiple epiphyseal dysplasia. Am J Med Genet A 132A: 181-184,2005.

48. Odman P: Hereditary enchondral dysostosis: twelve cases in three generations mainly with peripheral location. Acta Radio152: 97-113, 1959.

49. Paassilta P, Lohiniva 1, Annunen L et al: COL9A3: a third locus for multiple epiphyseal dysplasia. Am 1 Hum Genet 64: 1036-1044, 1999.

50. Ramachandran G, Mason D: Double-layered patella: marker for multiple epiphyseal dysplasia. Am J Orthop 33: 33-36,2004.

51. Ribbing S: Studien uber hereditare multiple epiphysenstorungen Acta Radiol Suppl 34:7-100, 1937.

52. Ribbing S: The hereditary multiple epiphyseal disturbance and its consequences for the aetiogenesis of local malacias; particularly the osteochondrosis dissecans. Acta Orthop Scand 24: 2860299, 1955.

53. Rimoin DL, Lachman RS: Epiphyseal dysplasias. In McKusick's Heritable Disorders of Connective Tissue, Beighton P Editor. St. Louis, Mosby 1993. 622-625.

54. Rubenstein JD, Christakis MS: Case 95: Fracture of double-layered patella in multiple epiphyseal dysplasia. Radiology 239: 911-913,2006.

55. Sebik A, Sebik F, Kutluay E et al: the orthopaedic aspects of multiple epiphyseal dysplasia. Int Orthop 22: 417-421, 1998.

56. Senee V, Vattem KM, Delepine M et al: Wolcott-Rallison syndrome: clinical, genetic and functional study of EIF2AK3 mutations and suggestion of genetic heterogeneity. Diabetes 53: 1876-1883,2004.

57. Sheffield EG: Double-layered patella in multiple epiphyseal dysplasia: a valuable clue in the diagnosis. J Pediatr Orthop 18: 123-128, 1998.

58. Thur J, Rosenberg K, Nitsche Pet al: Mutations in cartilage oligomeric matrix protein causing pseudoachondroplasia and multiple epiphyseal dysplasia affect binding of calcium and collagen I, II and IX. J BioI Chern 276: 6083-6092,2001.

59. Unger SL, Briggs MD, Holden P et al: Multiple epiphyseal dysplasia: radiographic abnormalities correlated with genotype. Pediatr Radiol 31: 10-18, 2001.

60. Unger S, Hecht JT: Psuedoachondroplasia and multiple epiphyseal dysplasia: new etiologic developments. Am J Med Genet 106: 244-250,2001.

61. Watanabe S: Clinical study of knee joints in familial multiple epiphyseal dysplasia. Knee 19: 41-46, 1993.

62. Wolcott CD, Rallison MV: Infancy-onset diabetes mellitus and multiple epiphyseal dysplasia. J Pediatr 80: 292-297, 1972.

63. Waugh W: Dysplasia epiphyseal is multiplex in three sisters. J Bone Joint Surg 34B: 82-87, 1952.

ABOUT THE AUTHORS

About Henry J. Mankin, MD

Dr. Henry J. Mankin was born and raised in Pittsburgh, PA and received his Undergraduate Degree and MD from the University. He and his recently deceased wife, Carole had three children, one of whom, Keith, is a co-author of the book.

In 1953, Henry served as a resident at the University of Chicago and then 1955, he became a medical physician in the Navy during the end of the Korean War. He served for two years and then took a residency in Orthopaedics at the Hospital for Joint Disease, completing his training in 1960, at which time he joined the staff at the University of Pittsburgh and served until 1966. He then returned to the Hospital for Joint Disease as Chief of Orthopaedics until 1972. At that point he became Chief of Orthopaedics at the Massachusetts General Hospital and the Edith M. Ashley Professor of Orthopaedics at Harvard Medical School. He retired as an active participant in 2002 but has remained on the faculty and continuing with research and education until the present time.

Dr. Mankin served in many capacities in organizations. He was President of the Orthopaedic Research Society, the American Board of Orthopaedic Surgery, the American Orthopaedic Association, Musculoskeletal Tumor Society and has been on the Dean's Visiting Committee at the University of Pittsburgh since 1992. He has served as an honorary member of the Argentine, Canadian, Japanese, Israeli, British, Thai, Australian and New Zealand Orthopaedic Societies and is an honorary member of the British Royal Society of Medicine.

Dr. Mankin has had continuous research funding from the National Institute of Health from 1967 to 2002 and has published two books and 700 articles in Journals, many related to biologic research on bone, cartilage structure and function and on connective tissue tumors and some genetic disorders.

About Keith P. Mankin, MD

Keith P. Mankin, MD is a pediatric orthopaedist in the Raleigh, NC area. He too was born in Pittsburgh and attended University of Pittsburgh Medical School. He trained in Boston, working alongside his father at the Massachusetts General Hospital before moving South. He is married to Julia Fielding, a radiologist at UNC-Chapel Hill, and the father of Cameron, an artist who contributed this book's cover.